A TRIUMPHANT VOYAGE
Great Achievements in Cardiology

A TRIUMPHANT VOYAGE
Great Achievements in Cardiology

That Have Led Us to Today and Toward Tomorrow

Riaz Haider

A Triumphant Voyage: Great Achievements in Cardiology That Have Led Us to Today — and Toward Tomorrow

ISBN: 978-0-578-66177-3
Library of Congress Control Number: 2020904619

Printed in the United States of America

Cover Artwork by Nadya Scheiner

The author acknowledges and expresses sincere appreciation to and for the sources of information used in developing this book. The majority of these sources are listed in the bibliography at the end of the book. In addition, the book uses photos and images from the U.S. National Library of Medicine and from various websites, publications, and so forth, obtained via web research; the author greatly appreciates the ability to use these.

In appreciation of the pioneers in cardiology,
and of the many individuals
whose shoulders they stood upon,
each working for the benefit of all humanity

A Triumphant Voyage: Great Achievements in Cardiology

Table of Contents

"The Moving Finger writes; and, having writ,
moves on: Nor all your Piety nor Wit
Shall lure it back to cancel half a Line,
Nor all your Tears wash out a word of it."

– Omar Khayyam

Author's Note

This book is a guide to the must-see stops while traveling through the landscape of scientific and medical discovery in cardiology. It is a journey through the past to today and toward tomorrow, with glimpses of great achievements by pioneers in science and medicine.

I worked in cardiology across nearly 5 decades, caring for patients, teaching, conducting research and more. As I reflect with deep satisfaction on my career, I wish to share with both the general reader and those with a medical background the compelling saga of a triumphant voyage of pioneering accomplishments leading up to cardiology in the 21st century.

Advancements are determined by ideas. We shall begin our travels with a brief look at some ideas which, through the adventure of human reason and the illuminating spark of thought, have lifted us to a higher life.

After discussing the beginnings, we shall go on to the turning point in cardiology that came in the 17th century with William Harvey, who discovered the function of the heart and the general circulation.

Nothing of note happened in medical understanding of the heart for more than 250 years after that … until the 20th century. That is when the remarkable march of progress took place — with great discoveries and advancements, of which 10 will be described.

This book is inspired by the ideas and teachings that many fine individuals have impressed upon my mind over the years. In addition, while numerous exceptional people have directly and indirectly influenced my work and my life, I would like to respectfully recognize here three notable physicians: Amir ud-Din, MD (Pakistan); John F. Goodwin, MD (UK); and W. Proctor Harvey, MD (U.S.) To each of them, I extend my deepest gratitude.

— *Riaz Haider*

"... they may learn and experience the noble strain in humanity and learn that man is capable of such great things that even the gods might be envious."

– John Little (Canadian writer)

I. Prologue

i. Ideas, Human Reason and Inventions

From an obscure age to our own time, ideas and innovations from great thinkers have determined history and lifted civilization to a higher life. These thinkers include philosophers, mathematicians and scientists. They glimpsed pivotal truths and each in their own way, had an enduring influence on humanity. There are numerous names that one might include, but we will mention here just a few of the many thinkers.

Confucius (551-479 BC) taught us a reliance on facts and reason as a guide to human affairs. The Chinese philosopher also emphasized a strong sense of ethics, morality and justice. He was the first to teach the Golden Rule of the Axial Age: *What you do not wish for yourself, do not do to others*. He believed in compassion and transcendent goodness as the highest values.

Aristotle (384-322 BC) of Greece greatly advanced the idea of logical, scientific thinking. While he was incorrect about the Earth's central position in the universe — in "*On the Heavens,*" he expressed the idea, based on mystical reasoning, that the Earth was the center of the universe and that the moon, the sun and the stars moved around the Earth in circular orbits — for many other reasons, the history of science is considered to have begun with him.

The Greco-Roman **Ptolemy (100-170 AD)**, through further elaboration of Aristotle's idea, proposed a complex cosmological model in which the Earth stood at the center surrounded by eight spheres carrying the moon, the sun, the stars and the then-known planets: Mercury, Venus, Mars, Jupiter and Saturn. What lay beyond the last sphere of the stars was not made clear, as that was not mankind's observable sphere at the time. (This model of the universe was adopted by the Christian Church as being in accordance with the scripture because it had the advantage of leaving room outside the sphere of fixed stars for Heaven and Hell.)

Avicenna/Ibn Sina (980-1037) of Persia/Central Asia was a great physician as well as a teacher, philosopher and literary man. The primacy of intellect and the fundamental concept of the mind-body relationship can be traced back to him. He propounded a method of examination and experimentation that was among the first principles of science, as described in his five encyclopedic works together called the *Canon of Medicine*. Avicenna remained a medical authority — recognized in the Middle East/Central Asia and across Europe — for many centuries and is now described as the father of early modern medicine.

"The knowledge of anything, since all things have causes, is not acquired or complete unless it is known by its causes. ... Therefore, in medicine we ought to know the causes of sickness and health."

Avicenna/Ibn Sina

Copernicus (1473-1543) of Royal Prussia, Poland proposed, initially anonymously for fear of being accused of atheistic blasphemy, a simpler, circular model where the sun was at the center of the planetary system. The accepted dogma of the church at the time regarded the Earth as God's footstool and home of His redeeming pilgrimage: Copernicus said that the Earth was a minor satellite of the sun. In 1543 he published his mathematical description of the heavens — *"De Revolutionibus Orbium Coelestium," On the Revolutions of Heavenly Orbs* — as a single system circulating around the sun. The book was well named, for no book has created a greater revolution.

In Italy, **Galileo (1564-1642)** defied the holy Christian establishment in defense of a theory, not his own but that of a dead man named Copernicus, that the Earth was not the center of the universe — because he believed the theory to be true. Galileo was imprisoned for his beliefs and died while still a prisoner in his house. **Johannes Kepler (1571-1630)** of Germany modified Copernicus' theory and suggested that the planets move not in circles but in ellipses (elongated circles). The predictions now matched the observations.

In England, physicist **Isaac Newton (1643-1727)** explained that the planets were made to orbit the sun by universal gravitational forces. He said that the idea of gravitation came to him *"occasioned by the fall of an apple."* Newton's book *Philosophia Naturalis Principia Mathematica* is probably the single most important work ever published in the physical sciences. Newton's theory of gravity was based on a model in which bodies attract each other by a gravitational force proportional to a quantity — mass — and inversely proportional to the square of the distance between them. The theory predicted the motion of the sun, the moon and the planets to a high degree of accuracy. His laws of motion and mechanics led to later practical advances.

Albert Einstein (1879-1955) is known for his services to theoretical physics, especially the theory of relativity and the development of quantum theory. He joined light to time, time to space, energy to matter, matter to space and space to gravitation and the forces of electricity and magnetism. The beginning of the universe was brought into the realm of science by American astronomer **Edwin Hubble (1889-1953)**, who made the landmark observation in 1929 that wherever you look, distant galaxies are rapidly moving away from us, as the universe is expanding. His work was pivotal to the development of an important idea: that the universe came into being billions of years ago via the 'Big Bang.'

Charles Darwin (1809-1882) of Great Britain taught us that the species that endure are not necessarily the strongest but those that are best suited to their environment. He set forth his theories in his book, *On the Origin of Species by Means of Natural Selection*. The theory of evolution by natural selection was the single most important scientific innovation of the 19th century. Evolution was seen as creator of the originality of the universe. (In spite of arguments to the contrary, many religious people have accepted the idea of evolution, believing that God's wisdom and divine guidance underlie the process.)

The ultimate journey is surely that of life on Earth: Some 3.5 billion years separate the first primordial microbes from the emergence of a species capable of creating the miracles of music, art, literature, architecture, science and more. A species capable of complex and abstract thinking. The truths proclaimed by Copernicus, Kepler, Newton, Darwin and others and by their great work did not influence the morals or manners of their age, nor was there any immediate tangible or practical benefit (that could be explained to the appreciation of an ordinary person): Their huge accomplishment was that their work advanced rational human thought, our greater knowledge and understanding.

ii. Beginnings – Heart and Circulatory System

Ancient Greece

Hippocrates II, of Kos (460-370 BC) was a Greek physician who, along with others, correctly identified the heart as having four cavities, although his descriptions were based on the myths and beliefs of his time. Galen (see below) perpetuated Hippocratic medicine, moving knowledge both forward and backward. (Note: The 'Hippocratic Oath,' taken by many medical graduates even today as they swear to uphold specific ethical standards, is deemed an early expression of medical ethics in the Western world.)

Galen (129-200 AD) appears to have been, after Hippocrates, the foremost among early Greek physicians. The early practice of medicine was dominated by Galen's writings, a central concept –

dating back to the ancient Greeks — being the four humors: black bile, yellow bile, blood and phlegm, an imbalance of which was thought to cause disease. Galen believed in three types of spirits: natural, vital and animal. He held that food was transported via the portal vein from the intestine to the liver, where it became blood adorned with the natural spirit, thus providing the body with nutrition and the ability to grow. Galen conceived of the circulatory system as ebb and flow, consisting of two components: arteries and veins joined together by invisible poles in the atrio-ventricular septum of the heart. He was an anatomist, experimental physiologist and highly successful physician. These attributes plus the Christian Church's general acceptance of his work help to explain his unshakable authority across almost 12 centuries, which held back the growth of knowledge about the physiologic basis of the heart's action and circulation of the blood.

The 13th Century

The first true challenge to Galen was posed by the early Muslim physicians. In the 13th century, the Arab physician **Ibn al-Nafis (1213-1288)**, for example, proposed the following regarding the lesser circulation, i.e., the pulmonary circulation: He believed that the blood passes through the vena arteriosa to the lung, spreads through its substance, mixes with air and becomes completely purified; then it passes through the arterio venosa to reach the left chamber of the heart. This thought was not only a learned man's brilliant insight, but a daring assault on the Galenic doctrine. The manuscript in which al-Nafis expressed his views was entitled *Commentary on the Anatomy of Avicenna's Canon*. Muslim translators state that it was this document that in due course initiated the pivotal anti-Galenic trend at the University of Padua, Italy.

The 15th – 17th Centuries

However, it was in the 15th through the 17th centuries in Europe that truly scientific medicine emerged, based on anatomical dissection and physiologic investigation. The anatomical drawings done by **Leonardo da Vinci (1452-1519)** depict the human body in detail. He noted the structure of the heart musculature and the aortic valve. That he did not go a step further to describe circulation of the blood is perhaps understandable, as with his many interests — in art, mathematics and other areas — he had many ideas to pursue.

Hieronymus Fabricius ab Aquapendente (1537-1619) continued the anatomic tradition at the University of Padua in Italy: While serving as professor of surgery, he would demonstrate anatomy by conducting dissections. His treatise on the valves of the veins appears to have stimulated a certain young man's interest and experimentation into the physiology of circulation: the youthful English physician **William Harvey (1578-1657)**.

In Fabricius, Harvey found a man to make his life's model. Fabricius was an enthusiastic teacher and investigator who also had other qualities that inspired Harvey's mind, those of generous sympathies and a keen sense of the wider responsibilities of his position; for example, Fabricius at his own expense built a new anatomical theater at the university.

iii. *The Beating Heart – a Puzzle!* Harvey Solves the Mystery

Harvey went on to make seminal contributions in anatomy and physiology. He is the first known physician to have described completely and in detail the systemic circulation and properties of the blood being pumped to the brain and body by the heart (although others before him had provided precursors of the theory).

Harvey's scientific thought and method of discovery answered the long-puzzling question: What does the heart beat for? His discovery of the motion of the heart and of blood circulation was central to the understanding of the body's workings. Harvey's triumph was, in fact, on the same high plane as the advancements of the great thinkers who preceded him.

His interest in the heart's valves, as noted, was stimulated by Fabricius' work showing the presence of venous valves in the human body. Harvey was eager to determine how structure related to action, function and purpose. Eventually, his work led to his discovering the blood's circulation, as Harvey proved that the blood passes from the periphery to the central parts.

William Harvey

He discovered that the venous valves are solely meant to prevent backward passage of the blood, a reflux, from the greater to the lesser veins.

Of his work, Harvey wrote:

> *"When I first gave my mind to vivisections as a means of discovering the motion and uses of the heart and sought to discover these from actual inspection and not from the writing of others, I found the task … so full of difficulties, that I was almost tempted to think … that the motion of the heart was only to be comprehended by God.*
>
> *At length, and by using greater and daily diligence, having frequent recourse to vivisections, employing a variety of animals for the purpose, and collating numerous observations, I thought that I had attained to the truth … discovered what I so much desired, both the motion and the use of the heart and arteries ..."*

Harvey's Understanding of the Heart and Circulation of the Blood

Harvey proved that the heart pumps the blood in a circular course. He observed that the actual quantity of blood, as measured after ejection from the heart, made it physically impossible for the blood to do other than return to the heart by the venous route — for otherwise the arteries in the body would explode under pressure. This revealed the venous valves' true function.

His discovery not only gave proof of circulation: It also showed his computation as the first instance of measurement in biologic investigation and thus gave an impetus to the rise of physiology.

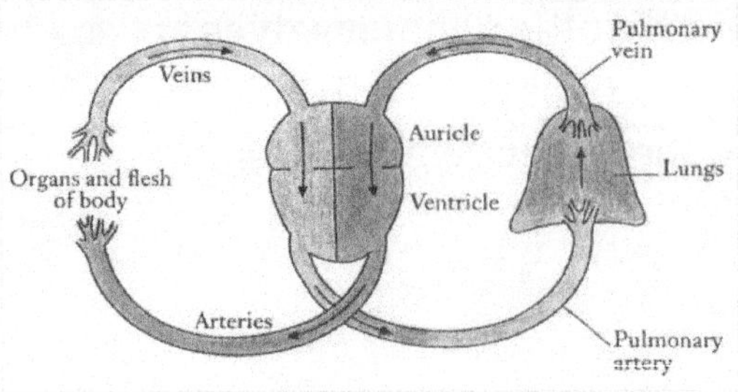

The heart and circulation of the blood, per Harvey's Understanding *(after Roy Porter)*

In 1628, Harvey in his monumental publication *Exercitatio Anatomica de Motu Cordis et Sanguinis in Animalibus,* generally referred to as *De Motu Cordis* [*Regarding the Motion of the Heart*], stated:

> *"It has been shown by reason and experiment that by the beat of the ventricles, blood flows through the lungs and it is pumped through the whole body. There it passes through pores in the flesh into the veins through which it returns from the periphery.... finally coming to the vena cava and right auricle ... It must then be concluded that the blood in the animal body moves around in a circle continuously, and that the action or function of the heart is to accomplish this by pumping. This is the only reason for the motion and beat of the heart."*

As William Osler would, centuries later, point out about Harvey and his work: For the first time, a great physiological problem was approached from the experimental side by a modern, scientific mind — a mind that could weigh evidence and allow conclusions to emerge naturally but firmly from observations based on thinking, devising and planning.

How Venous Valves Prevent Reflux of the Blood
William Harvey, in *De Motu Cordis* (1628)

"That this truth be made more apparent, let an arm be tied up above the elbow as if for phlebotomy ... At intervals in the course of the veins, especially in laboring people and those whose veins are large, certain knots and elevations (B, C, D, E, F) will be perceived ... These knots or risings are all formed by valves, which thus show themselves externally.

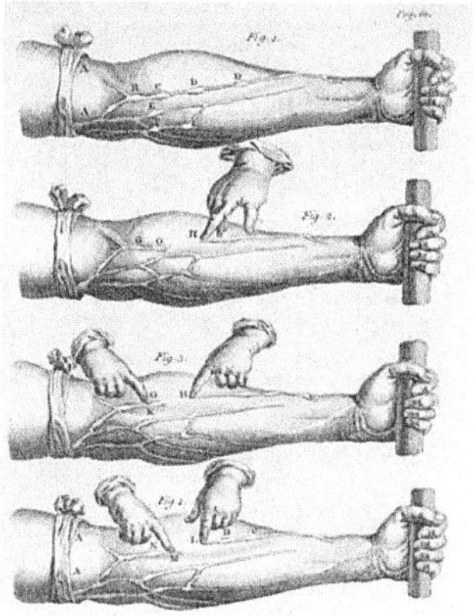

Per Harvey: Venous valves of the forearms

And now if you press the blood from the space above one of the valves, from H-O and keep the point of a finger upon the vein inferiorly, you will see influx of blood from above; the portion of the vein between the points of the finger and the valve O will be obliterated, yet will the vessel continue sufficiently distended above.

If you now apply a finger of the other hand on the distended part of the vein above the valve O and press downwards, you will find that you cannot force the blood through or beyond the valve. The finger first applied (H), removed, immediately the vein is filled from below. It would therefore appear that the function of the valves in the veins ... is ... to prevent all reflux of the blood that is passing over them." ◆

It has been said that *"he alone discovers, who proves,"* and in the matter of circulation of the blood, the credit belongs to Harvey. His *De Motu Cordis* dealt a death blow to the doctrine of spirits and humors that had prevailed since the days of Hippocrates and Galen, and Harvey thus changed the understanding of disease itself. It took great courage to depart from ancient doctrines and accepted opinion, as he did.

The true merit of Harvey's work lies not so much in the demonstration of the fact of circulation, as in the demonstration of the method. It is in this way that the *De Motu Cordis* marks the modern mind's break with the old traditions. From this we may date the beginning of scientific medicine. His single-minded pursuit of scientific truth by experiment opened the way in which science was done and led to collaboration of great minds across many different areas of science.

"Harvey sought the Truth, in Truth's own book —
The Creatures, which by God himself was writ —
And wisely thought it was fit,
Not to read comments only upon it,
But on the original itself look."

– Abraham Cowley (English poet)

WILLIAM HARVEY

Solved the Mystery: What Does the Heart Beat for?

Born in 1578 at Folkestone, England, William Harvey went on to attend Gonville & Caius College, Cambridge University, where he took his first degree. In the pursuit of medical education, Harvey next traveled to the University of Padua, in Italy, then the most celebrated school of medicine.

He was there when Galileo was achieving his fame at Padua. At the university, Harvey attended the anatomy lectures of the great Fabricius ab Aquapendente. In 1602, he received his medical degree at Padua and that same year was made a Doctor of Medicine at Cambridge.

He then settled in London to practice medicine, and he married the daughter of Lancelot Browne, who was Queen Elizabeth and King James I's physician. Harvey later became a Fellow of the (Royal) College of Physicians (RCP) and was subsequently appointed physician to the St. Bartholomew Hospital in London. Upon his appointment as Lumleian lecturer at the RCP in 1615, Harvey began his lectures in anatomy, in which he also described his work on the motion of the heart and blood.

In his lectures, he professed *"to learn and to teach anatomy not from books but from dissections, not from the position of philosophers, but from the fabric of nature."* During his career, he dissected more than 80 types of animals.

Harvey expounds conception of blood's circulation to King Charles I

Harvey had many illustrious patients, among them, King James I, to whom he became Physician Extraordinary. Upon that royal's death, Harvey became physician to the successor, King Charles I. The latter king shared Harvey's interest in movement of the heart and in medical investigations of the process of generation. In 1628, Harvey published his great work on the circulation of the blood, *Exercitatio Anatomica de Motu Cordis et Sanguinis in Animalibus*. He dedicated his book to King Charles I.

With his own funds, Harvey built and equipped a library for the College of Physicians, to which he made over his property in Essex, providing for a salary to the College Librarian and endowment of an annual oration to exhort the Fellows *"to search out and study the secrets of nature by way of experiment and also for the honor of the profession ..."* ◆

Later in the 17th century, the lung capillaries were discovered in the frog by **Marcello Malpighi**, thus solving the mystery of the passage of blood from arteries to veins. **Robert Hook** and **Richard Lower** by **1667** had shown that blood turned red when passing through the lungs of dogs. **John Mayow** in **1668** demonstrated the action of the lungs in taking something (thought of as "particles" by 17th century chemist **Robert Boyle**, now known for Boyle's Law) from the air into the blood to give it life; in doing so, Mayow may be credited with the discovery of oxygen.

For the next 250-plus years, there were no significant advances in medical understanding of the heart ... until the 20th century.

II. 20th Century: Great Achievements

At the dawn of the 20th century, the great achievements in cardiac medicine and cardiology began to take place — and the innovations steadily continued to appear.

Let us move forward in our journey.

"… the stories of the lives of the masters of medicine do much to stimulate our ambition and rouse our sympathies."

– Sir William Osler

Chapter 1
Electricity: Electrocardiography

Electrocardiography is the most common diagnostic procedure or test used today for heart ailments. According to some estimates, the test is performed 50 million times a year in the U.S., alone. The procedure is performed using an electrocardiograph machine that generates an electrocardiogram (ECG or EKG).

Heartbeats result from the discharge of electrical impulses in the heart that can be recorded as an electrocardiogram. In heart disease, there occur recognizable changes or abnormalities in the electrocardiogram. Thus, the human electrocardiogram is an invaluable modality in the diagnosis and management of heart disease.

Willem Einthoven (1860-1927), a professor of physiology in the Dutch town of Leiden, is recognized as the developer of the human electrocardiogram, a century ago. This essential, practical tool for diagnosing heart abnormalities and disease marked the birth of a new specialty: cardiology. Einthoven built on the earlier work done by many scientists in the field of electromagnetics and physiology. Before we discuss Einthoven's work, a brief history of electricity and of electrophysiology of the heart is presented in the following pages.

A Brief History of Electricity and of Electrophysiology of the Heart

Electrical and magnetic phenomena were certainly known in ancient times, but their explanations were based on mythical beliefs. It was only in the Middle Ages that a scientific inquiry of these phenomena began.

The ancients knew of two types of fish with electrical features that could induce electric shock: the electric catfish (Malapterurus electricus) and marbled electric ray/torpedo fish (Torpedo marmorata). Our ancient brethren also knew that amber — a fossilized (petrified) tree resin that occurs in various shades of yellow, orange, brown or white and is found nearly all over the world — becomes electrified when rubbed with cloth and can then attract or hold onto light objects. Due to this property, the Greeks gave it the name "electron." The Arabs knew of amber's special properties and called it *anbar*. It has also been stated that amber's electrical property was known in China during the time of Emperor Huang-Di (Qin Shi Huang, 259-210 BC).

The rock mineral magnetite, an iron ore, possesses magnetic properties. It was called lodestone in its later use as a compass for guiding mariners. The 13th century French crusader Petrus Peregrinus de Maricourt started an investigation of electrical phenomena in magnetism. He used lodestone as a model for his studies and developed the concept of the North Pole and South Pole and of polarization.

William Gilbert, one of the great English physicians of the late 16th century, throughout his life maintained a keen interest in electricity and magnetism. He carried out rigid experimentation. He recognized earth as a magnet, and he coined the terms "electric" and "charged body," later renamed as the "electric field."

Clarification and mathematical expression of various laws of electricity and magnetism were accomplished by **Henry Cavendish, Charles Augustin Coulomb** and **Simeon Denis Poisson.** Cavendish investigated the power of material to conduct static electricity and made measurements of electrical capacity. Although Cavendish anticipated the fundamental law of electrical flow, its final clarification was accomplished by **Georg Sinon Ohm**, a German physicist.

Italian scientist **Aloysio Luigi Galvani** in **1790** was able to move a dead frog's leg by electrical stimulation from a completed circuit connecting dissimilar metals. This led to the discovery that nervous action could be induced by artificial electrical phenomena — which marked the beginning of electrophysiology. Danish scientist **Hans Christian Oersted** observed that changes in electrical current could deflect a needle. The resulting measuring device became known as a "galvanometer," as a tribute to Galvani.

In **1856, Albert von Kölliker** and **Heinrich Müller** in Germany, while experimenting on two frogs, unknowingly allowed the sciatic nerve of one frog (with the gastrocnemius muscle still attached) to come into contact with the exposed beating heart of the other. They found the gastrocnemius muscle contracting in synchrony with the heartbeat of the exposed heart. They were the first to discover that the heart generated electricity; this showed that nerve transmission, muscle action and heartbeat are biological processes dependent upon electrical phenomena that can be measured and observed by electromagnetic means.

The Human Heart and the Evolution of Electrocardiography

Now, about the human heart. The electrical impulse that triggers human heartbeats originates from the sinoatrial node. This is the impulse-generating — *pacemaker* — tissue, a group of cells located in the heart's right atrium, or right upper chamber. If the sinoatrial node function is impaired due to disease, the pacemaker sites in the middle (nodal) or lower (ventricular) part of the heart take over pacemaker function but at lower heart rates.

Augustus Waller (1856-1922), a medical doctor and physiologist, recorded the first human "electrogram." He was the first to publish his findings and was a pioneer in acquiring extensive experience with this new diagnostic modality. His initial work on electrocardiography in a combined clinical and physiological setting was conducted at St. Mary's Hospital, London. It was there that he gave a public demonstration of his recording device.

Waller next published a paper in 1887 entitled *"A Demonstration on Man of the Electromotive Changes Accompanying the Heart Beat."* He showed that the current of the heart could be studied without opening the thorax (chest) of laboratory animals. These studies could be made by connecting the surface of the body of such animals with electrodes, to which a Lippmann (mercury) capillary electrometer could be attached. These connections later came to be known as leads. Waller also demonstrated that the currents of the heart in humans could be studied in the same way.

The capabilities of his devices were not, however, quite adequate for diagnostic purposes. Waller therefore concluded, *"I do not imagine that electrocardiography is likely to find any very extensive use in the hospital. It can at most be of rare and occasional use to afford a record of some rare anomaly of cardiac action."* However, Waller would be proven incorrect on this point, as shall be described shortly.

(L to R) Augustus Waller with Jimmy, his pet/companion dog and research partner; Waller's ECG set-up and his dog Jimmy

AUGUSTUS WALLER

Recorded the First "Electrogram"

(but not satisfactory for clinical use)

Augustus Desiré Waller was born in Paris in 1856. He was the son of eminent British neurophysiologist Augustus Volney Waller, known for his work on the degeneration of severed nerve fibers (Wallerian degeneration, named after him).

Following his father's death in 1870, the young man and his mother moved to Scotland, where he pursued his medical studies at the University of Aberdeen and University of Edinburgh. In 1883, he was appointed a lecturer in physiology at the London School of Medicine for Women.

Waller later married one of his students, Alice Mary Palmer — daughter of George Palmer, a member of Parliament and founder of the biscuit manufacturing company Huntley and Palmer.

Upon his marriage, his other students reacted as follows: On the blackboard, they wrote the inscription, *'Waller takes the biscuit.'* Waller walked into the classroom, saw the writing on the blackboard, and — not the least bit offended, and in good humor — added the words, *'and the tin, as well.'*

He was well known for his unusual personality, often using his pet/companion bulldog Jimmy in research demonstrations.

Besterman conveys the researcher-physician's unconventionality in vivid detail, as follows.

"Waller presented a different appearance from that of our physicians who were always soberly garbed in frock coats or morning coats and silk hats. He was a short stocky man, very light on his feet. His grey beard and double-breasted blue jacket made him exactly look like a skipper in the merchant navy.

He was invariably followed by his bulldog Jimmy, who had the distinction of having had a question asked about him by a member of Parliament in the House of Commons."

> *Question: "A bulldog was cruelly treated when a leather strap with sharp nails was wound around his neck and his feet were immersed in glass containing salts in solution and the jars in turn were connected with wires to the galvanometers. Such a cruel procedure should surely be dealt with under the 'Cruelty to Animals Act' of 1876?"*
>
> *Witty answer: "The dog in question [Jimmy] wore a leather collar ornamented with brass studs, and he was placed to stand in water to which some sodium chloride had been added, or in other words, common salt. If [you] my honourable friend had ever paddled the sea, [you] will fully appreciate the sensation obtained thereby from this simple pleasurable experience!"* ◆

Willem Einthoven in his laboratory

Willem Einthoven was able through further research work to transform the curious phenomenon of electrocardiography into an indispensable clinical recording device. In 1889, he attended the first International Congress of Physiology in Basel, Switzerland, where he saw a demonstration of how Waller recorded the human electrocardiogram with the (mercury) capillary electrometer. Einthoven spent the next 5 years working to further understand the heart's electrical activity.

Using a capillary electrometer of the type that Waller had introduced, Einthoven made important improvements in function and resolution. He was able to record (barely) satisfactory electrocardiograms through complex mathematical and physical approaches; however, he was unable to improve the capillary electrometer's diagnostic capabilities.

Thus, he began work on another instrument, the string galvanometer. The string galvanometer consisted of a thin, silver coated quartz filament that passed between two electromagnets. Electric current passed through the filament produced a movement that projected a shadow which was magnified and registered.

The string galvanometer provided quantifiable readings of higher quality than its precursor capillary electrometer. This was due to the thinness and minimal mass of the string, capable of recording the rapid deflections of cardiac electric potentials with a higher degree of fidelity — considerably higher than Waller's mercury capillary electrometer. It was also due to the ability of the operator to adjust the tension to regulate sensitivity and response time. Operation of the string galvanometer was simpler.

The production of clearer and well-defined curves on the electrocardiogram satisfied the requirements of the International Committee for the Unification of Physiologic Methods. This landmark discovery became of immense value in the diagnosis and management of heart disease. It opened new vistas in what subsequently became the specialty of cardiology and later the subspecialty of electrophysiology. Einthoven's insight and foresight enabled electrocardiography to emerge as a field of practical value and brought him accolades.

Many years later, Waller and Einthoven recognized each other's contributions to the evolution of electrocardiography. In his 1922 book, *The Electrical Action of the Human Heart,* Waller stated: *"Einthoven's galvanometer is to the capillary electrometer as a high is to low power of the microscope. It has opened a new chapter in the clinical study of heart disease."*

WILLEM EINTHOVEN
Discovered the Mechanism of the Electrocardiogram (ECG/EKG)
(satisfactory for clinical use)

Willem Einthoven was born in 1860 in Java, part of the Dutch East Indies (present day Indonesia), where his father was stationed as a military physician. The family returned to Holland upon his father's death.

Einthoven's interest in electrophysiology was stimulated in his medical school days by the renowned physiologist **Frans Donders**, who at the time was studying action currents of the heart. In 1885, Einthoven earned his MD and PhD degrees from the University of Utrecht in Holland (now, the Netherlands). He joined Leiden University as professor of Physiology in 1886. During his later years Einthoven dedicated his time to teaching and lecturing about electrocardiography.

In 1924, Einthoven was awarded the Nobel Prize in Physiology or Medicine *"for his discovery of the mechanism of the electrocardiogram."* A man of great integrity and honesty, he wanted to share half of the $40,000 prize money with his former lab assistant and valued collaborative partner, Van de Woerd. Einthoven searched far and wide for him. Learning that Van de Woerd had died, Einthoven prevailed upon his former colleague's two surviving sisters to accept "justifiably" half of the award money.

A vivid portrait of the man is also to be found in the obituary written by his colleague A.V. Hill:

> *"Einthoven was a master of physical technique. Despite his medical training, he was essentially a physicist, and the extraordinary value of his contributions to physiology and therefore to medicine emphasizes the way aptitude for physics can aid in the solution of physiological problems.*
>
> *Of the more personal side of Einthoven's life, one might write of the grace, beauty and simplicity of his character. He spoke with ease three languages as well as his own; he was a regular attendant at international gatherings; he threw all his weight on the side of good international relations in science.*
>
> *It was a wonderful thing to be his guest and to enjoy the delightful hospitality of his home. He invited me some years ago while we were attending a German Congress of physiologists at Tubingen, to stay with him in Leiden on my way back to England. … In 1924 we sailed together to America and … under the starlit sky, we walked on the upper deck discussing the random movements of electrons in conducting fibers, and other matters equally as strange.*
>
> *These personal details will emphasize what a loss his passing will be … to the good fellowship of physiologists throughout the world."* ◆

Development and evolution of the electrocardiogram

Right – Augustus Waller used the letters A, B, C and D to designate the deflections that he obtained with a (mercury) capillary electrometer.

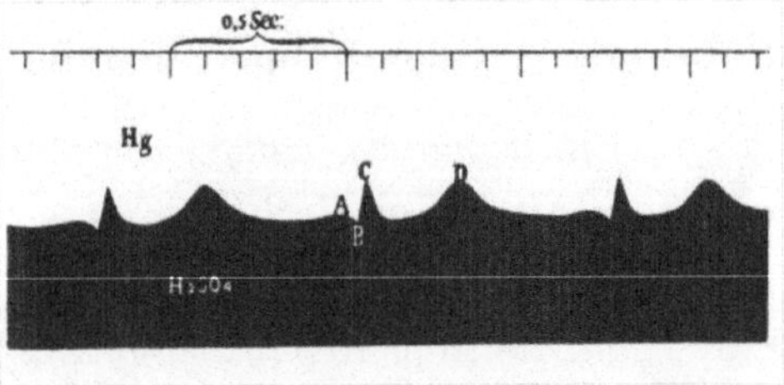

Right – Willem Einthoven substituted the letters P, Q, R, S and T for the deflections shown by his string galvanometer.

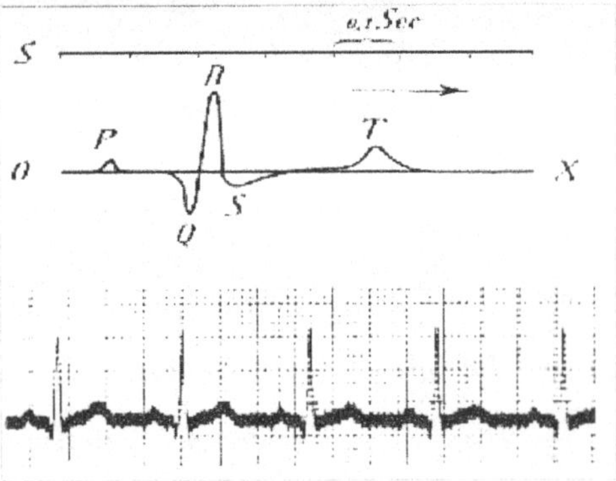

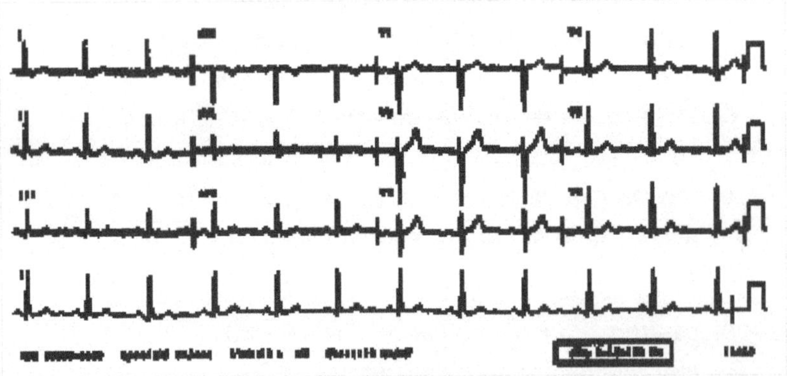

A normal human 12-lead electrocardiogram

British cardiologist **Thomas Lewis (1881-1945)** wrote of Einthoven: *"Honours, however, were to him a small recompense than was the knowledge of its benefits which his long and arduous work had conferred upon his fellow men."*

It was Lewis, at just 32 years of age, who wrote the first definitive textbook on clinical electrocardiography. The book was a product of the knowledge and experience gained in 4 short years, but it was written with a conviction that electrocardiography would become essential in the modern diagnosis of heart disease. In addition to the contributions of Waller and the pioneering work of Einthoven in discovering the mechanisms of the electrocardiogram, this era will be remembered for Lewis and for what later became the medical specialty called cardiology.

More than a hundred years after its discovery, the electrocardiogram remains an easily accessible central diagnostic tool in cardiology and its clinical practice. Such has been its enduring role.

Chapter 2
Audacity: Cardiac Catheterization

In early heart surgery, death on the operating table was not uncommon – and that was a psychologically devastating prospect and situation. When operating to close a hole in the heart, heart surgeons would find complex anatomical abnormalities that they were unprepared for and unable to fix by surgery. This was due to a lack of proper diagnosis preoperatively, as the only tools they and the cardiologists had available to them were a stethoscope, chest x-ray and an electrocardiogram.

Cannulation of the Heart/Cardiac Catheterization, Evolved in Three Periods

Cardiac surgeons needed precise knowledge of heart chambers, valves, inflow and outflow vessels and intracardiac pressures. This need was a major factor in the development and emergence of radiologically equipped cardiac catheterization laboratories dedicated to analysis of the heart.

"Cardiac catheterization shone a spotlight on dark places of the heart and illuminated with precision what had previously only been suspected. Cardiac catheterization made cardiac surgery possible and opened the way for later advances such as open-heart surgery, pacemakers and ultimately, cardiac transplantation."

– John F. Goodwin,
Clinical Cardiology (1993)

Cardiac catheterization, or cannulation of the heart, is a medical procedure in which a catheter, a long flexible tube, is introduced into a blood vessel in the arm or the thigh and advanced to the heart. The procedure has been an outstanding technological innovation that has enabled great advances in the diagnosis of congenital and acquired heart disease. It has also provided a safe method for highly innovative study of the structure and mechanical function of the heart in health and disease, and for treatment modalities.

Three Periods – Evolution of Cardiac Catheterization

Andre Cournand, a pioneer in the field of cardiac catheterization, describes its evolution in three periods: The first period is experimental, the second period is the demonstration in humans of its feasibility, and the third period is its elaboration and universal acceptance in the diagnosis and management of heart disease.

First Period: Measuring Intra-Cardiac Blood Pressure

Adhering to Newton's teachings in physical science and based on his belief in the order and balance of the universe, **Stephen Hales (1677-1761)** conducted studies on circulation. He showed intense interest in the quantitative aspects of pressure and flow in the living plants and animals. By direct cannulation of a mare's artery, Hales was able to measure arterial blood pressure and by correlating its peak blood pressure with heartbeat and the lowest level of blood pressure with its relaxed phase, he demonstrated the recognition of the concept of cardiac output and total peripheral resistance.

The first period really began, however, in **1844** with **Claude Bernard**, the first to perform the procedure in animals and naming it cardiac catheterization. He firmly believed that medical advances would occur with coordination of clinical observations and sound physiological studies. He went one step further by entering the heart chambers of animals — horses, sheep and dogs — through appropriate vascular channels. Thus, he was able to measure intracardiac blood pressure under a variety of physiological states.

By **1863, Jean Baptiste Chauveau** and **Etienne Jules Marey** based on their investigations had published the first systematic description and interpretation of intra-cardiac pressure recordings, achieving a milestone in cardiac physiology.

Second Period: Medicine's Strangest Story — and an Audacity

Werner Forssmann (1904-79) is known for challenging the medical taboo of his time — that one could not enter the living human heart to study it because such an event would be fatal. Forssmann tells the story in his book, *Experiments on Myself: Memories of a Surgeon in Germany.* From our perspective today, it is the story of a daring physician of the not too distant past who opened the way to the beating heart.

Forssmann had read that 75 years earlier, a Frenchman experimenting on a horse had successfully pushed a tube (a cannula, for cannulation) from a peripheral vessel all the way to the heart. Forssmann was convinced that this could be done in humans. In the

1920s, he explained his thoughts on the subject to Peter Schneider, the hospital chief physician. The common understanding at the time was that just like surgery on the heart, pushing a tube into the heart ran the risk of ventricular fibrillation and death. Schneider's response, therefore, was: *"I cannot possibly allow you to carry out such an experiment on a patient."* When Forssmann said that he would accept the risk by pushing a catheter in himself, Schneider was shocked by the young doctor's absurd and grandiose response, and countered, *"My 'no' is final and absolute."*

Forssmann carried out an initial experiment on a cadaver. He introduced a ureteral catheter through the cadaver's right arm vein and noted that he could easily advance the catheter to the right atrium of the heart. He then persuaded his colleagues to try the procedure on him. They were unsuccessful and refused to try again, fearful of possible adverse consequences. Undaunted, Forssmann made plans to carry out the procedure on himself. In 1929, he proceeded — with radiographic evidence. *(See story on following pages.)* However, due to unsubstantiated fears about the possible dangers of Forssmann's discovered technique of entry into the heart, the medical world had to wait 12 years for the next leap.

Werner Forssmann

WERNER FORSSMANN

Conducted First 'Cardiac Catheterization'— an Experimental Version, on Himself

Born in Berlin in 1904, Werner Forssmann began life in adversity — his mother raised him on her own after his father died in World War I (WWI) — and as an adult, worked with determination against the odds.

When Forssmann was a surgical resident at a hospital in Eberswalde, a forest town near Berlin, he sought to develop a technique for rapid, effective administration of drugs directly into the heart. He also saw in it new possibilities of evaluation of the heart and circulation in health and disease.

At the hospital in 1929, Forssmann courageously decided to conduct a 'cardiac catheterization' experiment on himself. First, he sought to win over the surgical nurse who kept the keys to the cabinet of sterile supplies — as she would be his unwitting accomplice. In his own words: "*I ... started to prowl around Nurse Gerda Ditzen like a sweet-toothed cat around the cream jug ...*" Charmed by Forssmann, she readily did his bidding and provided the supplies.

When the hospital staff was taking an afternoon nap, Forssmann proceeded: He exposed one of his own left arm veins and under fluoroscopy control, introduced a ureteral catheter and advanced it to the right atrial chamber of his heart. To obtain a permanent record, Forssmann walked to the x-ray department and had the x ray technician awaken the radiologist, Peter Romus, from his nap. Romus charged into the room and screamed at Forssmann to stop what appeared to be a rendezvous with death. Forssmann, however, carried on; he turned on the x-ray machine and recorded the event with a chest x-ray film. In this way, he provided proof of a safe entry into the heart — and, the transition from physiological experimentation to clinical application was assured.

His report appeared in Germany's premier medical journal *Deutsche Medizinische* in 1929. It caused a sensation in some medical circles, but reactions mostly ranged from indifference to intense scorn. Elated by his success, however, Forssmann approached Ernest Suerbruch, at the time an internationally esteemed authority on thoracic surgery, with the good news — hoping for its acceptance as a valid device. He was dismissed with a statement that Suerbruch ran a clinic, not a circus.

Undaunted and with his enthusiasm undiminished, Forssmann experimented again on himself. In 1931, he injected a radio opaque substance into his heart and had himself x-rayed (thereby laying the foundation of recording images of the heart, now called angiocardiography). But, he discovered that the dons of German cardiology were not welcoming to the Eberswalde 'troublemaker.'

He thus entered training as a urological surgeon. ◆

Third Period: Cardiac Catheterization Is Established as a Diagnostic Tool

Some years passed after Forssmann's discovery. Then, it was collaborative work done by **André F. Cournand (1895-1988)** and **Dickinson Richards (1895-1973)** at the Columbia University Chest Service at Bellevue Hospital, New York City, that made the world aware of the great possibilities inherent in cardiac catheterization. This was the third period, which saw cardiac catheterization established as a highly accurate and useful diagnostic tool in cardiology. As the two leaders in the field, Cournand and Richards further developed the technique, and greatly so.

Lawrence Henderson had proposed a hypothesis that lungs, heart and circulation form a single system of the exchange of respiratory gases between atmosphere and the tissues of the organism. Cournand and Richards, to validate the hypothesis, embarked upon the use of right heart catheterization to obtain mixed-venous blood so as to determine respiratory gas concentration.

Cournand described in detail how it happened (a native of Paris, France, he had immigrated to the U.S. in 1931):

> *"In 1936 [at Bellevue Hospital in New York] we considered whether the technique of right heart catheterization might provide help in obtaining the mixed-venous blood needed for determination of respiratory gas concentrations. In accordance with our general plan, I arranged to visit with Dr. Ameuille, one of my former teachers, who had completed pulmonary angiography in 100 patients in his hospital in Paris.*

The protocol of these cases convinced me of the safety of the technique. Dick Richards and I, in collaboration with Dr. Darling, then familiarized ourselves with the method [in animal experiments]. Meanwhile, George Wright … offered to advance urethral catheters under fluoroscopic control, into the right atrium of human cadavers, in order to measure the distance between the catheter tip and the landmark on the anterior chest wall on lateral chest x-rays; this datum would be needed to establish a reference point for measuring intra-cardiac pressures."

Cournand et al. published a report, "*Catheterization of Right Auricle in Man,*" that indicated the importance of cardiac catheterization as an accurate and reliable clinical tool. This paper codified the technique of catheterization and outlined a precise method for determination of cardiac output, intracavitary and pulmonary artery pressures. The authors provided irrefutable evidence supporting the validity of the Fick principle (used for measuring cardiac output) under basal conditions and after physical effort.

André Cournand

Cournand emphasized that scientific data should be interpreted not in isolation but in the full, proper context of the clinical problem to be solved. He firmly believed in the primacy of the patient — human, fearful and hopeful, seeking relief, help and reassurance. In a memorial to Cournand, who died in his 93rd year, J. Lequime, a former student of Cournand and later a professor of cardiology at the University of Brussels, summarized Cournand's view as to the primacy of the clinical aspect in all his research efforts.

Lequime quoted Cournand's own words:

> *"To be fruitful in the useful application of new methods requires not only a profound knowledge of the techniques used but also that of the patient under examination: the value of precise physiologic measurement is all the greater when the clinical study is more elaborated; before presenting his conclusions, the investigator must always ensure that he has explored in all its clinical aspects the problem which he proposes to solve."*

John McMichael at the Hammersmith Hospital and the Royal Postgraduate Medical School (now the Imperial College of Medicine, London) introduced cardiac catheterization in England. In addition, dismayed that orthodox medicine had banished Forssmann from the academic world, McMichael sought out the doctor, who was living in near-oblivion in Germany. McMichael helped to restore Forssmann to his rightful place in the evolution of cardiac catheterization.

Just as electrocardiography had done half a century earlier for electrical function of the heart, cardiac catheterization opened the way for elucidation of mechanical function of the heart. Forssmann's imagination and audacity found the technique to enter the human heart. Cournand and Richards' diligent pursuit of scientific investigation, generating profound results, brought meaning to it.

It was a giant step forward in solving the puzzle and the mystery surrounding the human heart. The discoveries related to cardiac catheterization made heart surgery possible and opened the way to pacemaker implantation, heart transplantation, coronary angiography and coronary angioplasty. Cournand and Richards — together with Forssmann — were awarded the Nobel Prize in Physiology or Medicine in 1956.

Chapter 3
Serendipity: Coronary Angiography

Changing Times, Changing Health Issues

With industrialization, population death rates and the causes of death have changed. Average life expectancy in the U.S. at the beginning of the 20th century was 47 years of age — whereas in the year 2000, it exceeded 75 years of age. This astounding advance is a particular reminder of how far we have come in our understanding of disease pathophysiology, diagnosis, treatment and prevention.

Until the early 1900s, infectious diseases were the major cause(s) of death. With the advent of antibiotics and vaccines, however, there was a steep decline in such deaths. This change began with the discovery of penicillin by (Sir) Alexander Fleming in 1928 at St. Mary's Hospital, London, and of penicillin's practical use by Howard Florey and Ernst Chain at the Royal Infirmary, Oxford in 1940 (followed by their landmark article in *The Lancet*). Fleming and Florey/Chain were jointly awarded the Nobel Prize in Physiology or Medicine in 1945, *"for the discovery of penicillin and its curative effect in various infectious diseases."*

The shift away from deaths due to infectious disease continued with the development of vaccines, such as the (Jonas) Salk vaccine against polio in 1955. Then: Around the mid-20th century, heart disease/coronary artery disease became a major cause of death.

Heart Disease/Coronary Artery Disease

With deaths from infectious disease vastly reduced by the mid-20th century, the average life expectancy of Americans reached 68 years. At this time, cancer and chronic disease predominated — and heart disease emerged as the leading cause of death. The shift was in part because Americans were living long enough to develop heart disease, which commonly presents itself from the fifth through the seventh decade of life. In addition, there was a general decrease in people's physical activity levels due to labor-saving devices and automation. There was also a change in food habits from a traditional low-fat agrarian diet toward a diet heavy in saturated fats and, as more recently recognized, in refined carbohydrates.

All these factors acting over decades on Americans were affecting their blood vessels, encouraging atherosclerosis, or hardening of the arteries (both coronary and peripheral). The recognition and management of heart disease became an urgent matter. Let us discuss how the development of coronary angiography changed this situation.

A Major Diagnostic Advance: Coronary Angiography

Coronary arteries supply blood rich in oxygen to the heart muscle, thus enabling it to perform its function. Coronary angiography is a special x-ray procedure for visualizing how well blood flows through the coronary arteries. A cardiac catheterization procedure is used to inject a radio opaque contrast material (a dye) using a flexible tube called a catheter into the coronary arteries.

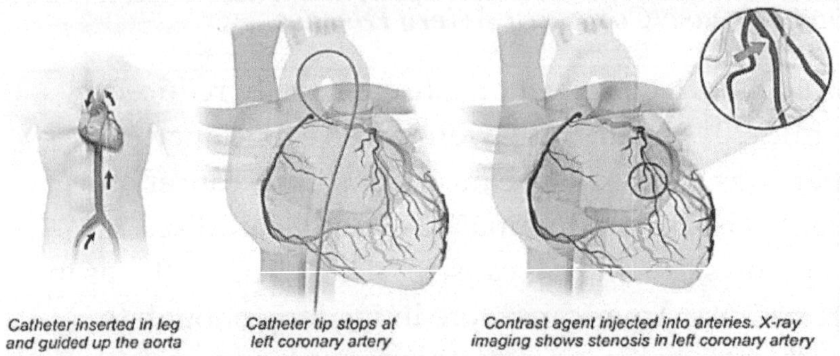

Coronary angiography

X-ray pictures are obtained while the dye is flowing in the lumen of the coronary arteries, thus outlining the lumen and revealing any area of blockage therein. Whereas history-taking and physical examination of a patient remain fundamental, and electrocardiography remains important, coronary angiography — by precisely delineating the location, extent and severity of coronary artery obstruction — emerged as a major advance in the diagnosis of coronary artery disease.

Slow-growing hardening (atherosclerosis) of the coronary arteries does not usually precipitate a major cardiac event. What does? It is the abrupt and catastrophic rupture of the atherosclerotic lipid (fat)-laden patch or plaque that develops within the inside lining of the coronary arteries and the exposure of substances that promote clot formation that produces *a blockage in the coronary arteries* — thus causing a sudden heart attack (acute myocardial infarction). Coronary angiography reveals the site and size of this blockage.

This major advance has been instrumental in the development of coronary artery bypass surgery as well as interventional cardiology.

A Great Turning Point: A Gift of Serendipity

In the late 1950s and early 1960s, x-ray equipment was still primitive, and catheters suitable for coronary angiography were not available. Furthermore, the risk from intracoronary injections was considered too high, with anginal pain, dangerous arrhythmias or even sudden death greatly feared.

Then, in 1958, **F. Mason Sones** (1918-85), a cardiologist at the Cleveland Clinic, transformed the situation. Sones had a patient who was a 26-six-year-old man with rheumatic mitral and aortic valve disease. After completing a left ventricular angiogram on the patient and to next perform an aortogram, Sones placed a closed-end catheter in the ascending aorta, just above the aortic valve. Next, he loaded the pressure injector with 50 ml of contrast material. To monitor the injection, Sones climbed down into the pit under the catheterization table, where the 11-inch image intensifier was located. He directed his cardiology fellow to fire the pressure injector.

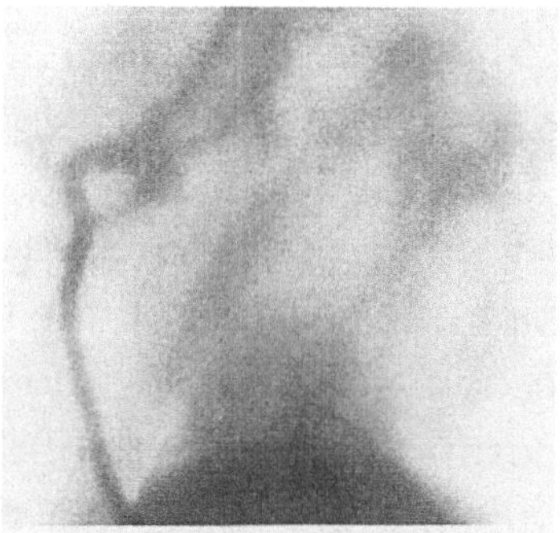

Right coronary artery (RCA) angiogram, left anterior oblique position, shows slippage of catheter (top left area) by Mason Sones

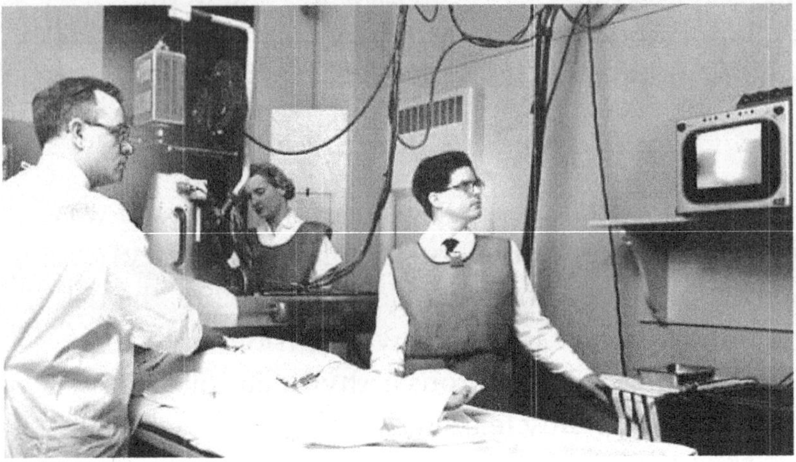

Mason Sones (l) conducts a cardiac catheterization procedure.

To his astonishment and horror, most of the contrast material was delivered directly into the right coronary artery — with the catheter whiplashed into its ostium, or opening. Fearing the worst, Sones leapt up from the pit and grabbed a scalpel, ready to cut the chest and perform open chest cardiac massage (a closed-chest variant for cardiac arrest would be introduced only later, in 1960, by William Kouwenhoven).

But to everyone's relief, the patient only had asystole, stoppage of beating of the heart, for approximately 5 seconds, and it was followed by sinus bradycardia, abnormal slow beating of the heart. Vigorous coughing by him and the injection of atropine sulfate by the team restored the patient's normal cardiac sinus rhythm. The patient recovered within a minute — and was perplexed to see Sones with a scalpel in hand, seemingly ready to 'attack' him!

Although performance of this selective arteriogram was not intentional, it was the first direct injection of a contrast agent into a coronary artery. It gave a transient but perfect image of the patient's right coronary artery. Rather than ignoring it as a difficult procedural event, Sones saw its potential. A few minutes later, he walked to his office and announced to his staff, *"We just revolutionized cardiology."*

The result of Sones' unintentional procedure was the first selective coronary arteriogram/angiogram. This serendipitous event — visualization of the coronary arteries — marked a new era in cardiology. (Note: Instances of serendipity in science are well known: for example, Greek mathematician-scientist Archimedes' discovery of buoyancy and Newton's discovery of gravity. But it takes a prepared mind to recognize the significance of a serendipitous event.) Sones understood, correctly, that a catheter could be inserted into a coronary artery and a contrast material could be injected into it, thus providing an incandescent image of the disease that was causing the heart attack.

As understanding the cause precedes the cure, thanks to Sones, the coronary arteriogram/angiogram may be considered the beginning of the end for coronary artery disease: Finding out what was happening inside those arteries, e.g., a clot or thrombus due to a blockage, has over time led to other advances in the field. With the newly designed catheters, Sones and collaborators were able to demonstrate that the new method of coronary angiography was safe, and with proper technique, yielded excellent visualization of the coronary arteries.

MASON SONES

Performed First Selective Coronary Angiogram — and Was an Irrepressible Character

Mason Sones was a coronary angiography pioneer.

He was also, reportedly, an irrepressible character who delighted in shocking people. Unpretentious and colorful in his language, he cursed like a sailor, drank and smoked. He even kept a lit cigarette, within sterile forceps, in his lab at the Cleveland Clinic.

His daughter Patricia (Sones) Wheat recalls:

> *"My father [was a man] small in stature, who was larger than life in person. My earliest memories are of playing in the corridors of the B-10 Cardiac Catheterization Laboratory … while my father reviewed films of patients. … Once, while watching a catheterization in the lab, I passed out. My last memory was of my father yelling at the patient, "Cough! Cough!! Cough, you #&&*#!!"…*
> *[It was only after waking up] in the emergency room, with a concussion, I found out that during a catheterization, if a patient were to go into fibrillation, coughing could sometimes restore normal rhythm to the heart."* * ◆

** When a person's heart stops, a physician may be able to restart the beating of the heart by a thump on the chest or by urging the person to cough vigorously.*

News of the results obtained at the Cleveland Clinic spread rapidly throughout the world and drew hundreds of cardiologists to visit Sones' laboratory, although it was not until 1962 that the first published results appeared. A major modification was then introduced by **Melvin P. Judkins**. His percutaneous (affected through the skin) transfemoral approach (the Judkins technique) used pre-bent coronary-seeking catheters.

The real breakthrough came in **1961** when Sones demonstrated angiographically that implantation of internal thoracic (mammary) artery into the left ventricular wall (the method of **[Arthur] Vineberg**) could provide effective perfusion of the myocardium.

This was followed by interventions such as bypass surgery and eventually, by percutaneous interventions such as coronary angioplasty as we know it today. Sones worked with a new catheter especially designed and made for him by the U.S. Catheter and Instrument Company. It had a pre-formed tape that permitted easy entry into the coronary ostia without completely obstructing the artery.

Sones' observations on deliberate selective opacification of individual coronary arteries in more than 1,020 patients are presented in his landmark paper, "Cine Coronary Arteriography," published in 1962 in *Modern Concepts of Cardiovascular Disease.*

The opening paragraph of Sones' paper outlined the need for this new technique:

> *"During the past four years, the technique of cine coronary arteriography has been developed in an effort to provide more objective and precise standard of diagnosis for human coronary artery disease. Heretofore, the diagnosis of coronary atherosclerosis has been primarily dependent on the physician's interpretation of the symptoms of distressed patients who described chest pain, and upon recognition of transient or chronic electrocardiographic changes which usually indicate the presence of myocardial ischemia or necrosis.*
>
> *Although conscientious, knowledgeable history-taking and electrocardiographic study require no apologies for their contributions to understanding, their limitations have been responsible, even in the hands of experts, for the production of iatrogenic disability on the one hand, and unjustified reassurance on the other, in a significant number of patients. A safe and dependable method for demonstrating the physical characteristics of the human coronary artery tree, which could be applied in any one phase of the natural history of coronary artery disease, was needed to supplement available diagnostic methods."*

Sones and his staff developed a unique reporting system. From cine film, certain representative frames were selected and reproduced as enlarged prints on photographic paper. The narrative description of each frame was presented in a semi-standardized manner.

Recently, this reporting system has regained interest because digital techniques make reproduction of angiographic pictures a great deal easier.

During the past decade, non-invasive methods for coronary angiography, i.e., magnetic resonance imaging (MRI) and multi-slice computed tomography angiography (MSCTA) have emerged. These technologies are still being improved, and the future will tell us if they can entirely replace coronary angiography.

Chapter 4
Cardiovascular Surgery

Four Major Eras

According to pioneering heart surgeon **Dwight Harken**, there have been four major eras of cardiovascular surgery. Let us discuss these, as we continue with our travels.

The First Era: A Period of Extra-Cardiac Surgery

With its roots in antiquity, the first era lasted to the end of 19th century. It consisted of surgery directed outside of the heart (extra-cardiac), toward the great vessels, pericardium, and stab wounds of the heart seen on the battlefield — and which needed to be treated with urgency and without the benefit of anesthesia.

Most of the vascular surgery during the greater part of this period was limited to traumatic lesions of peripheral arteries. Attempts at arterial suturing were doomed to failure because of infection and thrombosis.

"Surgeons must be very careful
When they take the knife!
Underneath their fine incisions
Stirs the Culprit — Life!"

– Emily Dickinson (American poet)

At New York's Rockefeller Institute, **Alexis Carrel (1873-1944)**, born in France, introduced a successful leak-proof technique and procedure for anastomosing, or surgically connecting, blood vessels without constricting the lumen. In *"recognition of his work on vascular suture and the transplantation of blood vessels and organs,"* Carrel in 1912 received the Nobel Prize in Physiology or Medicine.

The Second Era: Congenital Heart Disease Surgery and 'Blind' or Closed Heart Surgery for Heart Valves

The second era was the period of congenital heart disease surgery and of 'blind' or closed-heart surgery for the heart valves. In the 20th century's first half, **Maude Abbott (1869-1940)** in Canada and **Helen Taussig (1898-1986)** in the U.S. at different times played a crucial role in advancing cardiovascular surgery for congenital heart disease — ushering in the dawn of a new era of heart surgery.

They delineated the heart's anatomy and structure, and they conducted pathophysiological analysis, especially for cyanotic heart disease, of the abnormal function of the heart.

Abbott and Taussig both lived in an era when women faced blatant discrimination and had to overcome great hurdles to make professional progress. They thus stand out not only for their extraordinary achievements in medical science but also as leaders in women's empowerment. They were ahead of their time.

Maude Abbott and her influential book

Much of Abbott's life's work concerned the nature of heart defects and disorders, especially in newborn babies. This would ultimately lead to her being recognized as a world authority on congenital heart defects. Abbott studied medicine in Canada and then, after completing postgraduate studies in Europe, returned to Montreal. In 1897, she opened an independent clinic dedicated to treating women and children. She became an expert on cyanosis and heart murmur and did much first-hand research in pathology.

In 1898 she was appointed Assistant Curator of the Pathological/Medical Museum at McGill University. In a 1905 article for the *Journal of the American Medical Association*, she wrote: *"The organized [medical] museum is to general pathology what the autopsy room is to medicine, what the dissection room is to anatomy, what … traveling to see new countries is to the study of geography."*

Abbott was sent by McGill to Washington, DC to study the methods of the U.S. Army Medical Museum. During her trip, she met physician and pathologist (Sir) William Osler, who was at John Hopkins University and Hospital. He eventually became a mentor to her and subsequently asked her to write the section on congenital heart disease for his co-edited book, *System of Modern Medicine (1908)*. Osler described Abbott's chapter, based on 400 cases, as being *"far and away the very best thing ever written on the subject."*

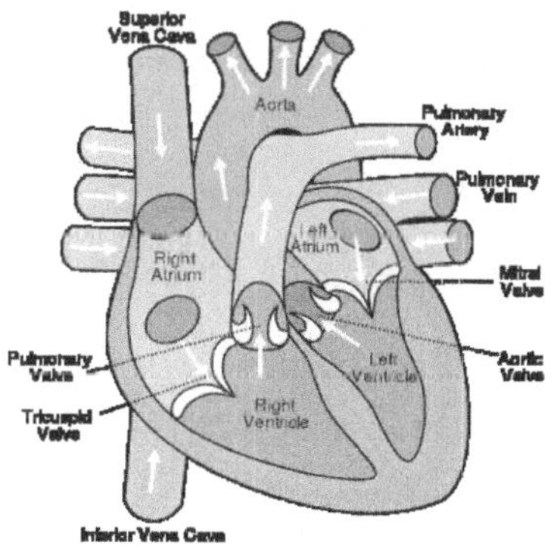

A normal human heart: Arrows indicate direction of blood flow among body, heart and lungs

Abbott's lifelong work in unlocking the secrets of congenital cardiac abnormalities was published in 1936 as the *Atlas of Congenital Cardiac Disease*. The atlas presented an analysis of 1,000 cases that she had personally studied. Her classification provided insights into the pathophysiologic pathway that enabled surgeons to devise rational surgical treatment to correct or ameliorate abnormal function.

Abbott's work, in fact, formed the basis of information about modern heart surgery. In the foreword to a later edition of Abbott's book, cardiologist Paul Dudley White wrote: *"Maude Abbott, fired by a spark from Osler, made the subject of congenital heart disease one of such general and widespread interest that we no longer regard it with either disdain or awe, as a mystery for the autopsy table alone to discover or to solve."*

MAUDE ABBOTT

Advanced Our Understanding of Congenital Heart Defects and Disease

Maude Abbott was born in East Quebec, Canada, in 1869 and raised by her grandmother, after tuberculosis claimed her mother when Abbott was an infant and her father abandoned the family.

McGill University's new policy of admitting women to its arts program enabled Abbott to study there, and she graduated in 1890 with high honors. However, McGill's Faculty of Medicine did not yet allow women to study medicine; despite her persistent efforts, she was refused admission. She instead enrolled at Bishop's College of Medicine, where she received her medical degree with the highest marks and won top academic prizes.

Following postgraduate studies in Europe, she returned to Montreal and established a clinic for women and children. In 1936 Abbott published the *Atlas of Congenital Cardiac Disease*, a book that had a powerful and lasting influence on heart surgery.

She was appointed Assistant Curator of the Pathological-Medical Museum at McGill and later founded what is now known as the International Academy of Pathology. To honor her key contributions, McGill established the Maude Abbott Medical Museum, a historical repository of materials. ◆

Earlier Attempts to Operate on the Human Heart

Modern cardiovascular surgery owes thanks also to a daring and successful procedure conducted by **Robert Gross** of Harvard Medical School and Boston Children's Hospital, in **1938**: He was the first to ligate and close a patent ductus arteriosus (PDA) successfully in a 7-year-old girl. The patent arterial duct connects two major arteries, the aorta and pulmonary artery, and is essential for fetal circulation but normally shrivels and closes after birth. When it persists, it strains the heart.

Gross had prepared himself for this bold operation by practicing it in the animal laboratory as well as in the post-mortem room. The indication for surgery was suggested to Gross by his colleague **John Hubbard**. The proposal was submitted to Kenneth Blackfan, who was chief of pediatrics but not a cardiologist. Blackfan suggested that they consult with respected cardiologist **Paul Dudley White**. After examining the patient, White said, *"Brave procedure, but obviously one that has been carefully considered. In my opinion, it should proceed as planned."*

Aware of his chief's reluctance regarding this untried operation, Gross decided to operate when Blackfan was away on vacation. The successful surgery generated a sensation around the world, and Gross and Hubbard reported their landmark success in 1939. The story goes that a few days after the operation, Gross encountered his chief, still on vacation, at the local club. When Blackfan asked Gross how affairs were at the hospital, Gross supposedly replied: *"Nothing unusual."*

It is ironic that not too many years later, with his fame now established, Gross showed an authoritarianism typical of many pioneers of that era, as he refused to help **Helen Taussig** when she proposed that he perform the "blue baby" operation on infants with cyanotic congenital heart disease/defects. *(Taussig's pioneering work on "blue babies" is described shortly.)*

The Birth of Early Heart Surgery

In the late 19th and early 20th centuries, no surgeon dared to operate on the heart. Theodor Billroth, a Prussian-Austrian authority on abdominal surgery but a fearful observer of the heart, in the late 19th century stated, *"A surgeon who tries to suture a heart wound deserves to lose the esteem of his colleagues."* In 1896, English surgeon Stephen Paget pompously proclaimed, *"Surgery of the heart has reached the limits set by Nature: no new discovery can overcome the difficulties that attend the wound of the heart."*

Nevertheless, American surgeon **Dwight Harken** attempted to operate on the heart. While working as a resident at the Royal Brompton Hospital in London, assisting pioneer thoracic surgeon A. Tudor Edwards, Harken had used the knife to treat patients with cancer and its extensions in the lungs and abdomen. He decided to shift his focus to attacking mechanical defects of the heart, a mechanical pump, by surgical means. At the time, infection affecting the heart valves, subacute

Dwight Harken

bacterial endocarditis, was a universally fatal disease. Harken had worked on experimental production of the disease in dogs and its surgical correction by excision, and he was now ready for its human trial. But Harken's research and plans were interrupted by World War II (WWII) and by two occurrences that influenced his path.

One was that the newly introduced drug penicillin rendered obsolete the indication of intracardiac surgery for infected heart valves. The second was that the Allied invasion of Europe took place while Harken, a member of the U.S. armed forces, was working as a thoracic surgeon in London, near the theater of war. It was the era of soldiers with artillery shell fragments in the heart.

It seemed to Harken that such intracardiac missiles could cause infection, embolization of the thrombus associated with the missile, or an aneurysm or effusion. These appeared to be indications for surgical removal of shell fragments from the heart. Harken firmly believed, in his words: *"You have to have a diagnosis that is absolute, condition that is incurable, and if you have any rational concept, you have the right to try."*

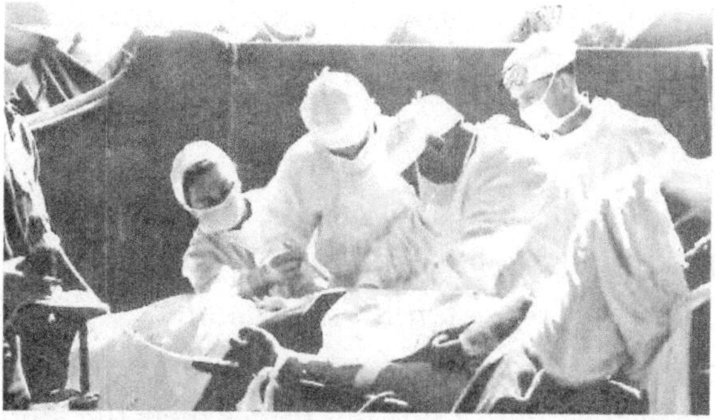

A surgical team operates on a soldier near the battlefield.

He consulted with available senior advisers American surgeon and professor Elliot Carr Cutler and British surgeon (Sir) Gordon Gordon-Taylor, and found genuine support for his suggestion that his technique for removing bacterial infections might lend itself to the removal of shell fragments. Professor Grey Turner, president of the Royal College of Surgeons, said to him, *"Yes, young man, all the reasons you marshal to support the removal of foreign bodies from these soldiers' hearts are prudent, and in addition, you must add the concern that a young soldier must have … that [he] harbours an unwelcome visitor in the citadel of his well-being."*

Harken found a way to safely take out shrapnel from the heart: He would cut into the wall of the beating heart, insert a finger to remove the shrapnel and close the incision by sutures.

In June 1944, a dying soldier with a gaping injury to his sternum and ribs was brought to his care. In the operating theater, upon opening the patient's chest — with assistants using retractors to widen the view — Harken noticed that a piece of shrapnel had penetrated the right ventricle of the heart. He succeeded in clamping on the shrapnel. He yanked … and a terrifying sequence of events ensued, as Harken describes in a letter to his wife.

> *"Then suddenly, with a pop, as if a champagne cork had been drawn, the fragment jumped out of the ventricle, forced by the pressure within the chamber … blood poured out in a torrent. I told the first and second assistants to cross the sutures and I put my finger over the awful leak.*

> *The torrent slowed, stopped, and with my finger in situ (in place over the wound) I took large needles swedged with silk and began passing through the heart muscle wall, under my finger and out the other side. With four of these in, I slowly removed my finger as one after the other was tied ... Blood pressure did drop, but the only moment of panic was when we discovered that one suture had gone through the glove on the finger that had stemmed the flood. I was sutured to the wall of the heart! We cut the glove, and I got loose ..." **

Harken had saved the life of a wounded soldier who otherwise would have died. It was exhilarating. It was a historic and revolutionary step forward. To this, we may date the birth of heart surgery.

It coincided with the Allied forces' invasion and landing along the beaches of the Normandy coast, a turning point of WWII, events that altered our times.

** Harken knew of the famous case from 1896 reported by German surgeon Ludwig Rehn: A man who had suffered a stab wound to his chest was brought unconscious to the hospital. After receiving an astute clinical diagnosis of massive pericardial effusion, the patient was quickly moved to the operating theater, where Rehn was the surgeon.*

Upon opening the chest, Rehn saw a heart trapped within the pericardial sac, distended with blood. He opened the sac with his scalpel, and blood gushed out. Rehn then noticed a half an inch tear in the wall of the right ventricle. His later description was as follows: "bleeding is controlled with finger pressure ... suture the heart wound tied in diastole ... bleeding diminished remarkably with the third suture ... heart rate and respiratory rate decreased, and pulse improved."

Rehn closed the chest, and 2 hours later, the patient was awake and resting comfortably. It was the first successful suture of a wound of the heart.

Harken became the first person to have repeated success in heart operations, as he removed shrapnel from the hearts of 130 wounded soldiers during WWII. There were 135 foreign bodies in or in relation to the heart and great vessels, and 17 were within the chambers of the heart. All were successfully removed, there was not a single death, and the first consistently successful elective intracardiac surgery was accomplished. Harken's stunning work was described and published in the *American Heart Journal* in 1946.

ENCOUNTERS WITH THE PIONEERS

A Tour

Led by Dr. Dwight Harken, USSR, 1982

"In 1982, I was part of a small group of cardiologists and cardiac surgeons who went on a tour of select clinical locations in what was then the USSR, or the Soviet Union. Our group was led by the pioneer of heart surgery, Dr. Dwight Harken.

We visited cardiac centers in Russia (Moscow), Tajikistan (Dushanbe) and Uzbekistan (Tashkent, Samarkand and Bukhara). Our tour provided ample time for me to discuss with Dr. Harken many topics related to the history of cardiac surgery.

I found his leadership qualities, congeniality, warmth and friendliness to be most endearing and admirable."

Congenital Heart Disease Surgery

Regarding modern cardiac surgery, American surgeon **Denton Cooley** writes, *"Harken's work helped overcome the notion that the heart cannot be surgically manipulated. It was a catalyst for the event that would mark the dawn of this era: … the Blalock-Taussig Shunt for the treatment of Tetralogy of Fallot, in 1944."*

Helen Taussig spent her career helping young children with debilitating heart conditions and is lauded as a pioneer in the development of pediatric cardiology as a specialty. In the 1930s and 1940s, during her time as physician in charge of the Pediatric Cardiac Clinic at Harriet Lane, a part of Johns Hopkins Hospital, she saw numerous patients with Tetralogy of Fallot. This was a common congenital heart malformation, appearing at birth or soon after and comprising the four developmental abnormalities described by various people, including Étienne-Louis Arthur Fallot (who coined the term "tetralogy" in 1888).

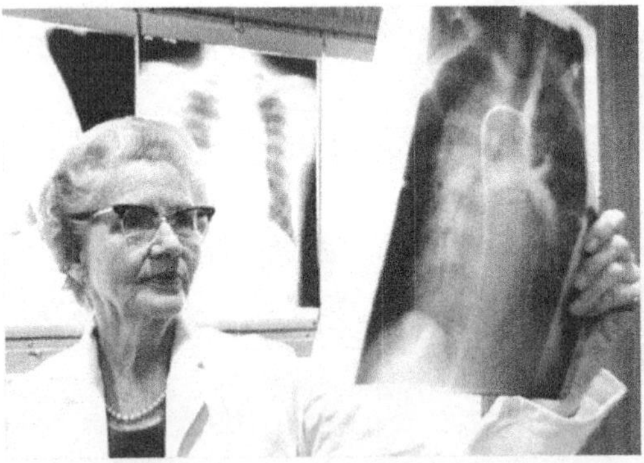

Helen Taussig

These infants had a bluish tinge to their skin (cyanosis). Due to severe narrowing of the pulmonary valve, there was insufficient blood flow to the lungs to get oxygenated, which resulted in the cyanosis; the infants were therefore sometimes referred to as "blue babies." Taussig had distinguished herself in pathophysiologic analysis and correctly realized that cyanosis in congenital lesions of the heart was dependent on pulmonary blood flow; the less the (oxygenated) blood flow, the deeper the cyanosis.

To improve the pulmonary circulation, she and her Johns Hopkins colleagues developed a procedure in which poorly oxygenated blood from the aorta could be directed to the pulmonary artery — thus bypassing the narrowed pulmonary valve. Referred to as aortopulmonary shunt, the "blue baby" operation was eventually named the Blalock-Taussig shunt, and is now also known as the Blalock-Thomas-Taussig shunt. *(See story later in this chapter.)*

In **1944** the first "blue baby" corrective surgery was performed by **Alfred Blalock,** at Johns Hopkins, on a 15-month-old infant girl. As reported, it was thrilling for all present to see the infant's previously blue-tinged skin color change to a pink glow. It was truly a physiological approach, restoring toward normal function; and although purely palliative, it was a remarkable breakthrough — again, until the advent of open-heart surgery, when corrective procedures could be implemented. Twenty years later, when Taussig gave a lecture at the scientific session of the American Heart Association, she reminisced about the earlier rebuff at the hands of Robert Gross, and portrayed vividly the salient features of the third operation and Blalock's surgical skills.

These are her words:

> *"Again, I am grateful to Harvard and to Dr. Robert Gross that while in his first flush of success of closure of patent ductus, in answer to my inquiry, he told me he had closed many a ductus but was not in the least interested in my suggestion of creating a ductus in a cyanotic child.*
>
> *So, I returned to Baltimore to bide my time until Dr. Alfred Blalock became surgeon-in-chief. The fact that the first three operations were successful attests to the brilliancy of Dr. Blalock's surgical skill. It was, however, at the end of the third operation that we saw the value of the operation. That operation was on an utterly miserable, small 6 years old boy that had a red cell count of 10 million and was no longer able to walk.*
>
> *When Dr. Blalock first removed the clamps, the blood welled up the child's chest. Dr. Blalock quickly controlled the hemorrhage and poured in plasma. Suddenly, Dr. Merrill Harmel cried, 'He is a lovely color now,' and I walked around to the head of the table and saw his lovely normal pink lips! The child woke up in the operating room and asked, 'Is the operation over'? When Dr. Blalock said, 'Yes', the child said, 'May I get up now?' From that moment on, he was a happy and active child."*

When Blalock and Taussig's paper appeared, there was a veritable odyssey of patients to Baltimore: Within 2 years, 500 patients underwent the procedure. A new era in the surgery of cyanotic congenital heart disease had dawned. Although the shunt procedure has now been replaced by other direct operations, the physiologic principles on which the procedure was based remain as sound as ever. With the advent of the heart-lung machine, total one stage correction of Tetralogy of Fallot became the method of choice, but indications for the palliative shunt remained.

HELEN TAUSSIG

Pioneer in Treating Congenital Heart Defects, Co-Invented the Blalock-Thomas-Taussig Shunt

Helen Brook Taussig was born in Cambridge, Massachusetts, where her father was a Professor of Economics at Harvard University (and at one time, an advisor to U.S. President Woodrow Wilson).

She earned her first degree in 1921 from the University of California, Berkeley. Taussig then sought to join Harvard as a student, but the university was not willing to admit women at that time (when she received an honorary degree from Harvard, years later, she reminded them of this.) So, she instead studied botany at Boston University. Taussig went on to study medicine at Johns Hopkins University School of Medicine in Baltimore, Maryland, one of the first medical schools to admit women. She qualified as a doctor in 1927 and later, became a pioneer in treating congenital heart defects.

She spent her entire, fruitful academic career at Johns Hopkins — she was eventually appointed Professor of Pediatrics — until retiring in 1963. Her writings spanned 60 years, as she believed that publication of research and clinical findings was a duty of the academic life.

In 1964, Taussig received the Medal of Freedom from U.S. President Lyndon Johnson. In 1965, she became the first woman and first pediatrician to be elected President of the American Heart Association. ◆

A Story of Medical Heroes and of the Blalock-Thomas-Taussig Shunt

Medical advancements are usually the combined result of numerous people's efforts, as well as the legacy of those who came before. Many individuals remain 'unsung' medical heroes. **Vivien Thomas,** who during his career at Johns Hopkins Hospital rose from surgical technician to director of the surgical research labs, was very nearly one of these.

Thomas over several decades worked closely with surgeon Alfred Blalock on developing operative techniques and designing surgical equipment. Thomas also did much fundamental research on his own. He was a key pioneer, along with Blalock and Helen Taussig, of the procedure for treating Tetralogy of Fallot.

It is said that Blalock was able to do just one practice shunt procedure in the animal model before performing the landmark "blue baby" corrective surgery in 1944: Thomas therefore stood with Blalock during the operation, advising him.

The Blalock-Taussig shunt procedure was later renamed the Blalock-Thomas-Taussig shunt, in belated recognition of Thomas' significant contributions. ◆

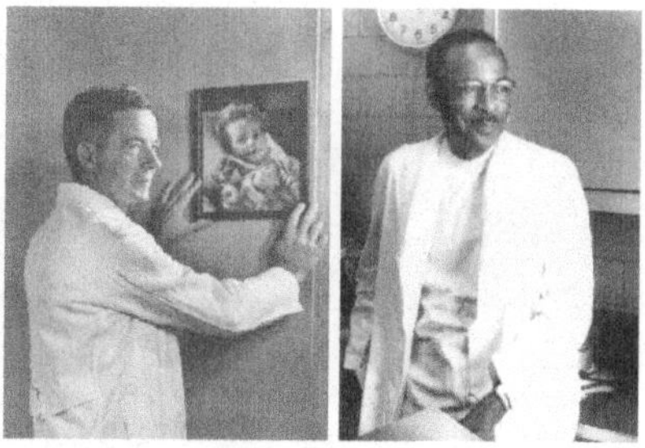

Alfred Blalock (l) and Vivien Thomas (r)

ENCOUNTERS WITH THE PIONEERS

Dr. Helen Taussig

at a Hospital in Birmingham, England, 1970

"In 1970, Dr. Helen Taussig, a pioneer in pediatric cardiology, visited the Children's Hospital in Birmingham, England while I was engaged in training there. She attended medical grand rounds with us.

At grand rounds, after the house physician presented the case of a patient with transposition of great vessels, I discussed the subject of cyanotic congenital heart disease. I then invited Dr. Taussig, at the time about 72, to share her comments. Her sharp intellect was impressive as she spoke of the importance of finding methods of providing oxygen to these oxygen-deprived children.

Dr. Taussig's words and presence were inspirational for me and for all in the room."

'Blind' or Closed-Heart Surgery

In 1925 in London, **Sir Henry Souttar** had for the first time used the trans-atrial approach in attempting to fracture the stenosed (constricted) mitral valve of a 19-year-old girl. She survived the closed mitral commissurotomy procedure; however, because of a lack of sophisticated anesthesia and antibiotics, the procedure did not gain acceptance.

In 1948 in Philadelphia, **Charles Bailey** — and later, **Russel Brock** in London — employed this approach and performed successful mitral valvotomy, but resulting mitral insufficiency remained a problem.

Harken, based on his laboratory experiments on dogs, found that damage to the posterior leaflet of the mitral valve was better tolerated. Based on this concept, he performed wedge resections of the fused mitral commissures with a valvotome and labelled the procedure a valvoplasty. The results with the aortic valve were not as successful. The techniques for correction of mitral regurgitation and aortic valve abnormalities did not advance sufficiently until open-heart surgery commenced.

The Third Era: Open-Heart Surgery

Until the mid-1950s, most pediatric heart operations were palliative extracardiac procedures performed on the closed heart. The challenge was to safely operate inside the heart to perform definitive intracardiac repair, or open-heart surgery.

What was needed was a method for interrupting blood flow during the intracardiac procedure. Hypothermia was one of the earlier methods tried, either placing the patient in a tub of ice water or cooling them with ice packs. Once the temperature was lowered to 26 C degrees, blood flow to the heart could be interrupted by placing a snare on the inferior or superior vena cava. But to prevent cerebral complications, the operation had to be completed within 8 to 10 minutes. The time limitation and danger of cerebral embolism made it clear that a more dependable method was needed, and preferably, quite soon.

Cross-Circulation and Open-Heart Surgery

C. Walton Lillehei (1918-99) at the University of Minnesota performed the first truly successful open-heart operation in 1954 — using cross-circulation, or the use of donor blood. He did so after first having an intuition: A mother's body could support the fetus, so why could another person's body not support the child?

Lillehei and his vision had their detractors; Cecil Watson, chief of medicine at the hospital, said, *"For the first time in history, a surgeon may have 200% mortality."* The mortality rate of open-heart surgery was high, and many surgeons despaired of ever being able to correct complex intracardiac defects. This situation changed in 1954 when, despite his detractors, Lillehei and his associates used controlled cross-circulation to correct a ventricular septal defect in a 20-month-old infant. The boy's anesthetized father served as the oxygenator.

C. Walton Lillehei

Blood flow was routed from the patient's caval system to the father's femoral (thigh) vein and lungs, where it was oxygenated and then returned to the patient's carotid artery. The cardiac defect was repaired with a total pump time of 19 minutes. Over the ensuing 15 months, Lillehei operated on 45 patients, most of them younger than 2 years old, with otherwise irreparable complex interventricular defects.

Although cross-circulation was a major advance, it was not adopted for widespread use because it posed risks to the donor. Nevertheless, this method paved the way for the open-heart surgery era. That same year, Lillehei and others introduced the first clinically successful bubble oxygenator, which remained the standard for extracorporeal circulation until the late 1970s. Surgeons could now operate on a variety of cardiac conditions under direct observation — a major advance.

Lillehei was both successful and one of cardiac surgery's great innovators. For their work, he and his associates in 1955 received the Albert Lasker Clinical Medical Research Award, America's most prestigious biomedical prize, for *"Advances in cardiac surgery — open-heart surgery — making possible more direct and safer approaches to the heart."*

The Heart-Lung Machine

Heart surgeon **John H. Gibbon (1903-73)**, who hailed from a long line of doctors, is considered the father of the heart-lung machine. In 1931, he assisted in a hazardous operation to remove a massive blood clot from the pulmonary artery of a female patient whose respirations had ceased and whose blood pressure could not be obtained. Despite the rapidity of the procedure, the patient could not be revived.

It occurred to Gibbon that the patient's life might have been saved if some of the blue blood in her veins could have been continuously withdrawn into an extracorporeal blood circuit, exposed to an atmosphere of oxygen and then returned to her by way of systemic artery in a central direction. Therefore, some of the patient's cardiovascular functions might have been temporarily performed by the extracorporeal blood circuit, providing more time while the massive embolus was surgically removed.

John and Mary Gibbon, with model II of their heart-lung machine

Gibbon provided a critical experiment in 1937 when he demonstrated that the cardiovascular function of cats could be maintained by a heart-lung machine, a cardiopulmonary bypass pump, during the clamping of the pulmonary artery for prolonged periods. He recognized that the most difficult part of the extracorporeal circuit would be the construction of an artificial lung. But he had an idea.

While a professor of surgery and head of surgical research at Jefferson Medical College in Philadelphia, Gibbon went to the IBM Corporation building in New York and succeeded in meeting with Thomas Watson, Sr., chairman of the board. Gibbon explained that he needed engineering help from IBM – and it was provided to him. From then on, he and IBM had a very productive and rewarding association. Gibbon and his research partner (and wife) **Mary Hopkinson Gibbon (1905-86)** – their colleagues fondly referred to them as "Jack and Mary"– worked over a long period on developing a screen oxygenator. This resulted in a greatly improved heart-lung machine.

How the heart-lung machine works:

The machine oxygenates venous blood and pumps it throughout the arterial system, via the ascending aorta or common femoral artery.

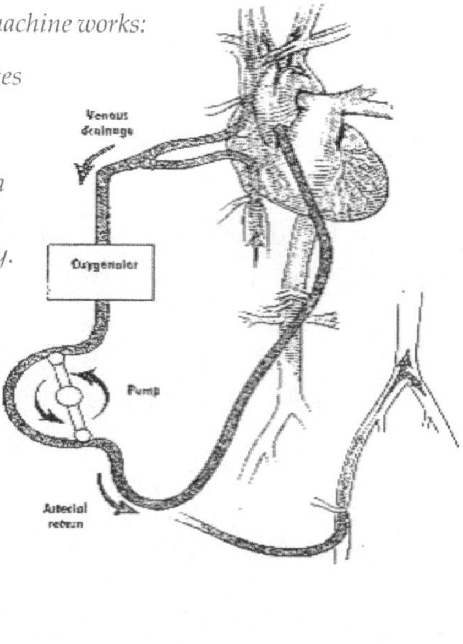

In 1953, using the enhanced machine, Gibbon successfully operated on an 18-year-old girl with heart failure. She had a large atrial septal defect, a hole in the wall separating the two upper chambers of the heart. The patient made a full recovery.

This was the first successful open-heart operation in the world using the heart-lung machine. It was also the first of many important, successful academic-industrial collaborations in cardiology. The heart-lung machine provided additional protection in the preservation of vital organs while enabling open-heart surgery, beginning with congenital heart disease and then heart valve repair.

Valvular Heart Disease: Valve Repair and Replacement

Infected and thus damaged heart valves — resistant to the newly arrived antibiotics and therefore resulting in abnormal heart function and shortened human lives — remained a challenge. The real breakthrough for valvular heart disease was total valve replacement, introduced in the early 1960s almost simultaneously by two individuals: Dwight Harken, mentioned earlier, who used a double caged ball and valve prosthesis, and **Albert Starr**, who used a caged ball and seal valve.

With the advent of open-heart surgery, Harken and Starr were pioneers in the implantation of artificial mitral and aortic valves.

Coronary Artery Disease and Revascularization Surgery.

With the accomplishment of major advances in the treatment of congenital heart disease and a decline in demand for surgery for valvular heart disease — along with a decline in rheumatic fever and rheumatic heart disease, due to the advent of antibiotics — a new challenge arose: coronary artery disease, which developed into almost an epidemic after WWII.

The key to tackling coronary artery disease was restoration of perfusion to the ischemic (blood-deprived) heart. Russian cardiac surgeon **Vasilli Kolenov** in **1964** performed the first internal mammary artery-coronary artery anastomosis (connection) to bypass coronary artery obstruction. But his innovations were not accepted by the medical community of those times.

In **1967, René Favaloro (1923-2000)**, who had trained in cardiovascular surgery at the Cleveland Clinic, performed the first coronary artery bypass operation, on a 51-year-old woman. With a segment of saphenous vein, the proximal and totally occluded right coronary artery was connected to the distal segment. A repeat cine coronary angiogram showed an excellent reconstruction.

The promising techniques of Favaloro were adopted immediately by **Dudley Johnson**. He extended the knowledge to the left coronary system and blazed the trail for multiple bypasses. Coronary artery bypass graft (CABG) then became the procedure of choice.

RENÉ FAVALORO

Carried Out First Coronary Artery Bypass Surgery

Born in 1923 in La Plata, Argentina, to Italian immigrants of modest means, René Favaloro graduated first in his medical school class during WWII. But due to his political differences with President Juan Peron's regime, he was denied an academic path. He therefore moved to the countryside and worked there as a surgeon for 12 years. Yet he dreamed of serving his nation better, by learning advanced techniques in cardiac surgery.

Thus, he came to the Cleveland Clinic in the U.S. Starting out as a laboratory technician during the day, he studied cine coronary angiograms with Mason Sones in the evenings. He then obtained a cardiac surgical fellow position and continued to progress from there. Favaloro ultimately turned out to be a star cardiac surgeon — in 1967, carrying out the first coronary artery bypass surgery. ◆

René Favaloro (l) with Mason Sones

George Green of New York later championed the idea, borrowed from **Robert Gertz**, of using internal mammary artery freed from the chest wall and connecting its free end directly to the side of the coronary artery just beyond the obstruction — which proved to be a distinct advance.

The era of revascularization surgery was made possible through the introduction of cine coronary arteriography/angiography by Mason Sones, described earlier, and by the development of the heart-lung machine by John Gibbon.

The Fourth Era: Heart Replacement Surgery

Cardiac Transplantation and the Artificial Heart

Heart transplantation as a treatment for end-stage heart failure was achieved by knowledge derived from numerous past experimental works. Important research was conducted early on by Alexis Carrel and Charles C. Guthrie (1905) and in the 1930s – 50s, by Frank Mann, Emanuel Marcus and Aldo A. Luisada, W.G. Downie and Vladimir Demikhov.

It was the foundational work done subsequently by cardiac surgeon **Norman E. Shumway (1923-2006)**, once a trainee of the great Walton Lillihei, that later allowed for human cardiac transplantation. In **1960**, at Stanford Hospital (of the Stanford University School of Medicine), **Shumway** and **Richard Lower** reported the first successful canine orthotopic cardiac transplant.

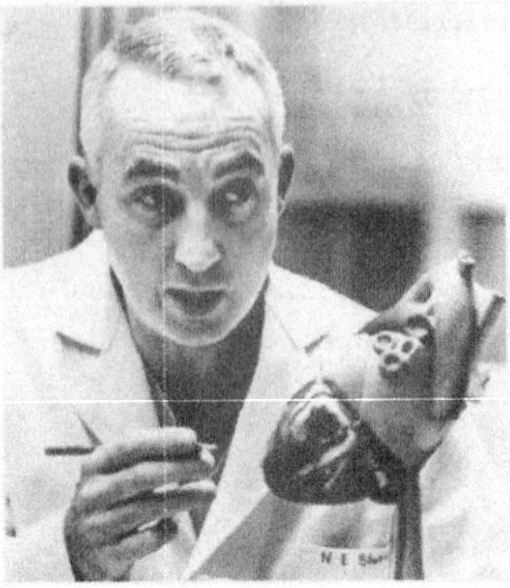

Norman Shumway, with a model of the human heart

Shumway was then ready to perform human heart transplantation. But two factors prevented him. The first obstacle was that in the U.S. in those days, brain death, while recognized as irreversible, was not yet part of the legal definition of death; death was defined as the absence of a heartbeat and of respiration. Removal of a beating heart from a (brain-) dead person was not permissible in that medico-legal environment — and waiting for cessation of the heartbeat and respiration for an indefinite period presented coordinational difficulties. The second obstacle that Shumway faced at this time was the likely rejection of the transplanted heart by the recipient's body, as the immunosuppressive drugs were as yet to be discovered.

In **1967, Christiaan Barnard**, a surgeon in Cape Town, South Africa who had also trained under Lillihei, used the Shumway technique to perform the first human heart transplantation — which enthralled the world. Barnard and his team obtained the donor heart from a woman in her mid-twenties who had sustained a severe head injury in a car accident.

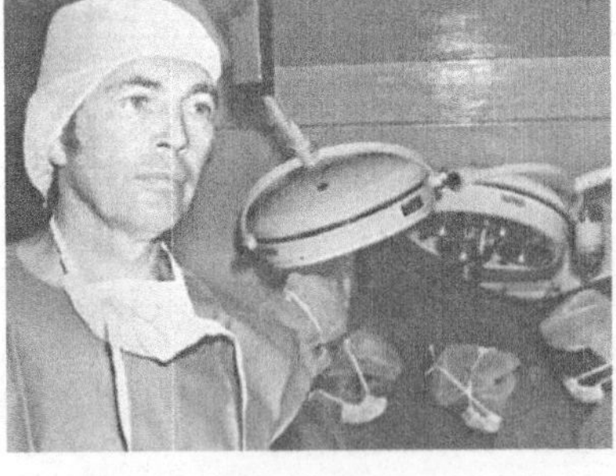

Christiaan Barnard

Shortly after her arrival at the hospital, the woman was declared brain-dead. Barnard then proceeded to carry out the transplant procedure on the recipient, a 57-year-old man with end-stage heart failure. Barnard performed only nine more heart transplants in the next 6 years, and he never publicly acknowledged Shumway's groundbreaking research work that had made Barnard's own achievements possible.

Meanwhile, Shumway persevered with his work and also corrected many flaws in the transplantation system, such as in the selection of compatible donors and recipients, organ preservation, and a schedule of heart biopsies. In **1968**, he conducted the first heart transplant in the U.S. and later led a team that performed the first successful heart-lung transplant.

He was also the first to use an immunosuppressive drug (cyclosporine) to prevent rejection and extend survival of the transplanted heart. Immunosuppressive protocols have, over time, greatly helped to improve transplantation results — reducing the rate of rejection and complications and increasing survival rates. For his groundbreaking work, Shumway is considered the father of cardiac transplantation.

But following Shumway's successes, another problem soon became apparent: Far too few donor hearts were available for the many patients who were heart transplant candidates. This discrepancy has persisted; in recent decades, only around 2,000 transplantations have been performed each year in the U.S., leaving many patients in need.

Development of Mechanical Support

The shortage of donor hearts impelled a search for mechanical support to serve as a bridge to transplantation or as a permanent treatment itself.

Michael DeBakey (1908-2008) pioneered development of the artificial heart, and the first clinical application took place in **1960** by **Denton Cooley (1920-2016),** mainly as a bridge to transplantation. Others approached the matter in different ways. In **1984** an artificial heart was permanently implanted in a human patient by **William DeVries** and colleagues. There was, however, a difficult post-operative course.

Presently, in an increasing number of patients, the goal of using a totally artificial heart has given way to treatment with a left ventricular assist device (LVAD), discussed later in this book. These are most commonly used as a bridge to transplant and as a destination therapy in patients with end-stage heart failure, especially those who fail to respond to medical therapy. More recent models of the LVAD have offered a good quality of life for some time. Many of these devices are now undergoing long-term evaluation.

Minimally Invasive Procedures — Uniting Surgeons and Cardiologists

A new era is dawning for performing minimally invasive heart surgery. Surgeons no longer need to cut through the breast bone (a sternotomy) but may now operate between the chest bones, which can result in less pain, less blood loss and quicker recovery, thus shortening the hospital stay for many people. In addition, hybrid operating rooms unite the surgical suite and cardiac catheterization lab, enabling surgeons and cardiologists to perform multiple procedures in one setting.

The minimally invasive procedures being conducted today include: single and multi-vessel coronary artery bypass surgery to improve blood flow to the heart and reduce chest pain, valve repair and replacement, surgery to correct atrial fibrillation, implantation of epicardial pacing wires to improve pumping action of the heart in heart failure, and treatment of congenital heart disease such as atrial septal defect.

New developments in cardiovascular surgery include robotically assisted heart surgery. Also called closed-chest surgery, this is a type of minimally invasive heart surgery in which the surgeon uses a specially designed computer console to control surgical instruments on robotic arms.

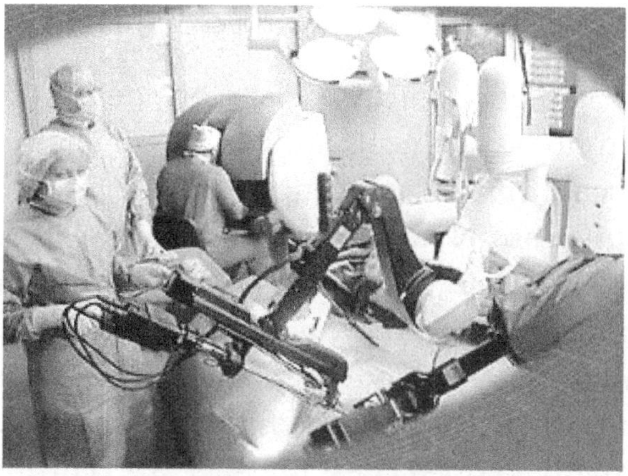

Cardiovascular surgery conducted using robotic equipment techniques, with lead surgeon (back) at computer console

Chapter 5
The Coronary Care Unit

The advent of the coronary care unit (CCU) also called the cardiac intensive care unit (CICU), has been described as the single most important advance in treating acute myocardial infarction, or heart attack. The CCU is a hospital unit staffed and equipped to treat patients with serious heart problems such as coronary thrombosis, i.e., a blood clot in the coronary artery that blocks blood flow, which may lead to damage to the heart muscle or acute myocardial infarction.

Allan Burns (1781-1813) in his classic 1809 work *Observations of Diseases of the Heart* had written a chapter entitled "Diseases of the coronary arteries and on syncope anginosa": *"Where however, the cessation of vital action is very complete, and continues long, we ought to inflate the lungs and pass electric shocks to the chest"* in the treatment of cardiac arrest. But for long afterward, nothing practical in that realm was accomplished.

In the mid-1950s, most cardiologists considered their main concern to be diagnosis and treatment of rheumatic heart disease and congenital heart disease — and not of myocardial infarction. To treat the latter, patients were sedated, placed on bedrest, administered stool softeners and moved away from the nurses' station. Anticoagulants were used, and adrenaline was employed in situations of shock. For cardiac arrest, treatment involved open chest massage, i.e., cutting open the chest and manually massaging the heart.

But the situation gradually began to change. In Boston in **1956, Paul Zoll (1911-99)** of Harvard Medical School and Beth Israel Hospital introduced external cardiac defibrillation, delivering a therapeutic dose of electricity to the heart with a device called a defibrillator. Working with technical collaborators, he also developed a way to display the heart's electrical activity on an oscilloscope screen, including registering each heartbeat with an audible signal and sounding an alarm at the onset of cardiac arrest. This ultimately led to cardiac monitors within modern cardiac care units.

In **1960**, work done by **William Kouwenhoven (1886-1975)**, a professor of electrical engineering at Johns Hopkins, showed a combination of the following to be effective in restoring cardiac function in patients with ventricular fibrillation: mouth-to-mouth breathing, sternal compression or closed-chest cardiac massage, and closed-chest electric cardiac defibrillation. The process was later named cardiopulmonary resuscitation, or CPR, and many people view Kouwenhoven as the father of CPR.

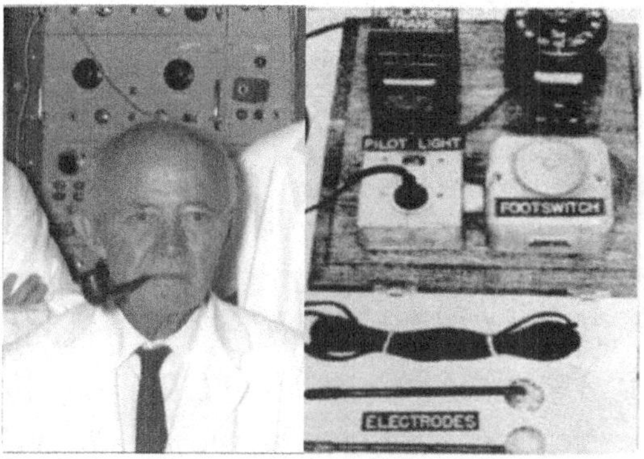

(L to R) William Kouwenhoven, in the lab; early defibrillator he helped develop

This major advance triggered the interest in providing intensive care for acute myocardial infarction. In 1973, when Kouwenhoven was presented the Albert Lasker Clinical Medical Research Award, the citation read, *"We salute you, Dr. Kouwenhoven, for your ageless genius ..."* Upon Kouwenhoven's death, the *New York Times* obituary on him observed that he had *"developed the basic cardiac treatment devices and procedures used worldwide."*

In **1961, Desmond Julian**, while working at the Royal Infirmary in Edinburgh, Scotland, articulated the concept of the coronary care unit. He described his vision of the CCU, based on four advances: defibrillators, pacemakers, heart rhythm TV monitors and CPR. His interest in coronary artery disease appears to have been stimulated when his father suffered a nonfatal myocardial infarction in 1954.

Later, while a fellow in Cardiology at the Peter Bent Brigham Hospital in Boston, Julian learned of Zoll's work and saw a demonstration at the Massachusetts General Hospital of an ECG machine that was triggered by the onset of a patient's cardiac arrhythmia. Based on his experience with CPR, upon his return to Edinburgh, Julian wrote: *"It became very clear that the potential for [CPR] was great but could not be realized because of the inherent delays when patients with myocardial infarction were scattered throughout the hospital, when there were very few trained in the techniques of CPR and when there was dearth of appropriate apparatus."*

Desmond Julian

Julian presented the first description of a CCU to the British Thoracic Society. He also wrote the following, published in *The Lancet* in 1961 and entitled, "Treatment of Cardiac Arrest in Acute Myocardial Infarction":

> *Many cases of cardiac arrest associated with acute myocardial ischemia could be treated successfully if all medical, nursing and auxiliary staff were trained in closed chest cardiac massage and if the cardiac rhythm of patients with acute myocardial infarction were monitored by electrocardiogram linked to an alarm system. All wards admitting patients with acute myocardial infarction should have a system capable of sounding the alarm at the onset of an important rhythm change and recording the rhythm automatically on ECG.*
> *The provision of the apparatus should not be prohibitively expensive if these patients were moved to a special intensive care unit. Such units should be staffed by suitably experienced people throughout the 24 hours."*

Julian relocated to Sydney, Australia and opened the first coronary/cardiac care unit there.

Today's Coronary Care Unit, or Cardiac Intensive Care Unit

The work done by both Julian and Kouwenhoven moved things forward considerably. Kouwenhoven's work eventually stimulated the interest of cardiologist **Hughes Day** of Kansas City, Kansas, who introduced to hospitals a mobile crash cart equipped with a defibrillator and external pacemaker. But the initial results were not promising because patients with acute myocardial infarction were scattered over different wards of the hospital. Day thus concluded that the patients should be monitored in a special environment, suitable when needed for immediate CPR — for this, he coined the term "coronary care unit," which was initiated at Bethany Medical Center in 1962. A few years later, in 1967, **Thomas Killip** and **John Kimball** reported their experience with 250 patients with acute myocardial infarction: Treated in a coronary care unit, the patients showed significantly better survival rates.

Frank Pantridge with his life-saving mobile defibrillator

Frank Pantridge (1916-2004), working in Belfast, Ireland, pioneered the use of a portable/mobile defibrillator and of mobile coronary care to shorten the time between occurrence of a heart attack and initiation of treatment. This was in view of the fact that, except in some cases of sudden death, the window of opportunity is limited to just a few first hours after heart attack; hence, the importance of the time factor. His achievement was a significant one.

Yevgeniy (Evgeny) Chazov in Moscow introduced the use of fibrinolytic therapy in acute myocardial infarction. He showed how this therapy leads to *"rapid relief of pain, less cardiac failure, less rise of blood transaminase reflecting the lessening of the damaged heart cells caused by a heart attack and more rapid signs of ECG healing" [signs of healing as reflected on an electrocardiogram].*

Yevgeniy Chazov introduced fibrinolytic therapy for heart attack

There had been initial attention given to arrhythmia detection; although important, it was overemphasized. It soon became clear that many patients who were saved from the electrical complications of a heart attack survived with varying degrees of lingering damage to the heart muscle.

The contributions made by Pantridge and Chazov in the prompt and appropriate treatment of a heart attack — when time is of essence to limit eventual infarct size and to prevent its further extension — while initially received with skepticism, have been proven correct with the passage of time.

Chapter 6
Preventive Cardiology

Prior to the middle of the 20th century, a heart attack was largely seen as a tragic inevitability that could rarely be predicted or prevented. However, this thinking underwent a revolution starting in **1944**, thanks to **Paul Dudley White (1886-1973).**

Among his many achievements, White may be best remembered as the father of modern cardiology and the founder of preventive cardiology. He pioneered the concept of preventing heart disease/coronary artery disease, during his tenure at Harvard Medical School and the Massachusetts General Hospital (MGH), and later as chief consultant to the National Heart Institute in Maryland (now, the National Heart, Lung and Blood Institute, part of the U.S. National Institutes of Health).

White advocated control of hypertension (high blood pressure) and obesity, opposed cigarette smoking and allowed for moderate use of alcohol. His general philosophy for a healthy way of living included three main elements: optimism, regular physical activity and work, with an emphasis on physical activity. White said, *"Walking is probably the best exercise [for the average person] because it is easy for anyone to accomplish and easy to grade from the slowest, shortest walk to the most rapid and longest."*

He was also a staunch proponent of vigorous exercise, and disagreed with medical opinion of that era that physical exertion could damage the heart. (In those days, after a heart attack, patients were often prescribed several months of bed rest.) Bicycling was one of his personal favorite forms of exercise. White was, in fact, inducted into the U.S. Bicycling Hall of Fame as a contributor who enhanced the sport of cycling through his lifelong efforts. He also emphasized the value of physical effort as an antidote to anxiety.

Paul Dudley White on his bicycle

White believed that having hope and a positive attitude toward life could help to maintain health and prevent illness. Above all, he believed that work was good for both mental and physical health, so he disavowed the practice of early retirement. White was one of the first to recognize the important role of lifestyle in the causation of coronary heart disease, and that it could occur even in young men. He was also a leader in emphasizing the critical role of cardiac rehabilitation for patients after myocardial infarction, now firmly established as standard practice.

The primary and secondary prevention of coronary artery disease (CAD) has contributed to almost one half of the dramatic 70 percent decline in [age-adjusted] CAD deaths.

PAUL DUDLEY WHITE
Father of Modern Cardiology, Founder of Preventive Cardiology

Paul Dudley White was born in 1886 in Roxbury, Massachusetts. As a student, he excelled in academics and sports. He often accompanied his father, a family practitioner, on hospital rounds and house calls (in a horse and buggy).

White graduated with honors from Harvard College and received his MD degree in 1911 from Harvard Medical School. He became a house officer at the Massachusetts General Hospital (MGH) that same year. In 1924 he married Helen Reid, whom he met while giving a lecture at Smith College Training School for Social Work. They subsequently adopted two children.

White traveled to England on a Harvard University traveling fellowship and studied under cardiologist Sir Thomas Lewis in 1913 at the University College Hospital, London. While there, he learned about a promising new diagnostic tool, the electrocardiography (ECG) machine, first developed by Einthoven a decade earlier. White brought an ECG machine back to the U.S. and was one of the first to use it here.

The London experience reportedly helped to shape his career, as did his sister's early death from rheumatic heart disease and his father's death at the age of 71 from coronary artery disease.

White joined the Harvard Medical School faculty and was from then on engaged in both clinical research and practical prevention of heart disease. White was one of six cardiologists who in the early 1920s founded the American Heart Association. In 1931, he published the classic text, *Heart Disease*. He remained on the Harvard faculty until 1948 and as emeritus professor until 1956. After a considerable period as chief of the MGH's cardiac unit, he departed in 1948.

Named president of the International Society of Cardiology, White also devoted time to developing the National Heart Institute, following the adoption of the National Heart Act signed into law in 1948 by U.S. President Harry Truman.

White delivered the very best care to all his patients. He was painstaking in taking a patient's medical history and examining them. His key message to his fellow doctors was a simple and humble but powerful one: *"Listen to what the patient can tell you. It may be more important than anything else you do."*

While most of his patients were ordinary folk, he also counted many prominent individuals among his patients. After U.S. President Dwight Eisenhower had a heart attack in 1955, White was appointed as his cardiologist. Following recovery from a heart attack, U.S. Senator Lyndon B. Johnson returned to the U.S. Senate only at White's encouragement — eventually going on to become the nation's 36th president. ◆

ENCOUNTERS WITH THE PIONEERS

Dr. Paul Dudley White at a Lecture in Boston, Massachusetts, 1969

*"I had the privilege of meeting Dr. Paul Dudley White, a pioneer in the prevention of heart disease, in 1969 while I was giving an invited lecture at the Massachusetts General Hospital (MGH), on "Health and Longevity in Hunza, Pakistan." * As I was speaking, I noticed Dr. White (by now, retired) seated in the back row. After I'd finished, he came up to the front and met me.*

We briefly engaged in conversation. Before we shook hands to say goodbye, Dr. White gave me a reprint of an American Heart Journal article (1964) that he'd co-authored, on the longevity of the Hunza people (his co-author/protégé Edward G. Toomey had traveled to the northern Pakistan mountain valley with a battery-powered ECG machine to collect data).

I cherish the memory of my encounter with Dr. White."

** As an assistant professor of medicine at Fatima Jinnah Medical College in Lahore, Pakistan, in 1965 — after completing my medical training in Pakistan and England — I led a team of researchers to the Hunza valley, noted for its populace's long life spans. Upon our return, we wrote and published our report.*

The Framingham Heart Study and Heart Disease Risk Factors

It was due to White's influential advocacy that the National Heart Institute (today, the National Heart, Lung and Blood Institute) in 1948 initiated the now world-renowned Framingham Heart Study, the first prospective population-based cohort study that focused on the risk factors for and causes of heart disease.

W.B. (William) Kannel (1923-2011) joined the Framingham study in the early 1950s, serving as its director from 1966 to 1979. Afterward he was a principal investigator on the study and remained involved with it until nearly the end of his life at age 87 in 2011.

"This type of study is a waiting game," Kannel said in an interview with PBS Television in 2006. *"We make measurements of the characteristics of the people we are following and then wait for them to get sick or not get sick."*

Over time, the Framingham Heart Study identified the major risk factors for coronary heart disease/ cardiovascular disease, to include obesity, hypertension, smoking and lack of exercise, as well as elevated cholesterol, and diabetes. While common knowledge today, these factors were noteworthy at the time. Smoking, for instance, had been known to cause pulmonary ailments like lung cancer and emphysema but had not previously been implicated in cardiovascular problems.

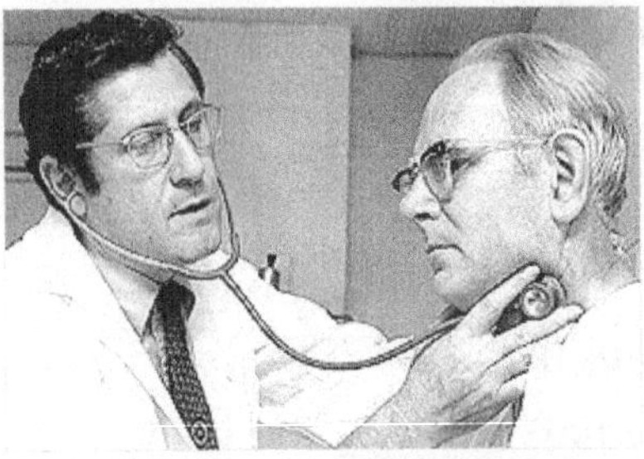

W.B. (William) Kannel

Furthermore, the study established that cardiovascular disease most often resulted from a web of interdependencies, or various risk factors acting in concert, rather than from one factor acting alone, as had commonly been believed.

Gathering and analyzing data on so many aspects of patients' medical and social lives was no small feat in the study's early days, as Kannel later noted.

"We had something like 80 variables to apply to this task and no computers, no copy machines," Kannel said, during the PBS interview. *"We were supposed to do this all by hand using carbon paper and electric typewriters and an abacus for counting and doing statistical analysis. We had a primitive punch card apparatus that did counting and sorting. The machine that did this was as big as an upright piano. It clanged away for 8 hours to count and sort what a computer could now do in 2 seconds."*

The Framingham study's findings are credited with altering the way doctors, patients and the public at large think about cardiovascular disease. We can thank the study, in part, for today's heightened awareness of the importance of behavioral and lifestyle changes — such as healthful eating, exercising and quitting smoking — in reducing disease risk and promoting good health.

"I think our surviving cohort have been wonderful and deserve all the credit they can get," said Kannel in the interview. *"It is now a contest actually between the senior investigators and the cohort to see who will survive longer."* He added, *"I suspect they [the cohort] will."*

Kannel lived a long life but was survived by some of the original cohort of men and women of the Framingham Heart Study. Now in its third generation, the study follows more than 14,000 people — among them, a few disease-defying, hardy members of the original research group.

The Framingham Heart Study provided information that has been crucial to the recognition and management of atherosclerosis (hardening of the coronary arteries), its causation and complications. It was also one of the first major studies in this field that included women as research participants.

THE FRAMINGHAM HEART STUDY
A Milestone in Cardiology Research, a Model for Future Studies

The Framingham Heart Study started in 1948 and continues to this day. Following an initial group of 5,000 individuals from Framingham, Massachusetts for their entire lives, the study asked two questions that were new for that time: For whom, exactly, was cardiovascular disease inevitable? Additionally, did cardiovascular disease have to be quite as inevitable as it was?

From its inception, the Framingham study was unusual. For one thing, epidemiological studies had until then been involved in infectious disease research. For another, where earlier clinical research had looked at patients who had already suffered heart attacks or strokes, this study admitted only healthy adults.

Subjects were given extensive physical/medical examinations, including interviews, every 2 years. Should one of them later have a heart attack or stroke, its underlying causes would already have been identified and well documented.

Thus, the Framingham study was a milestone in the history of cardiology and served as a model for many other longitudinal cohort studies. Those who initiated this study will be remembered as pioneers in preventive cardiology.

Interesting facts about the Framingham Heart Study:

~ The study is considered one of the most influential research endeavors in the history of medicine. It is how we know that smoking, high cholesterol and high blood pressure are major risk factors for heart disease. The findings have informed how doctors estimate patients' long-term risk of heart attack.

~ Administrators enrolled a second generation of participants, the original subjects' adult children and their spouses, in 1971. Still more people joined in 1994, 2002 and 2003, including new participants from more diverse backgrounds. The study's multigenerational nature has allowed doctors to better investigate the role of genetics in cardiovascular disease.

~ The study has led to the publication of 1,200-plus articles in major medical journals. ◆

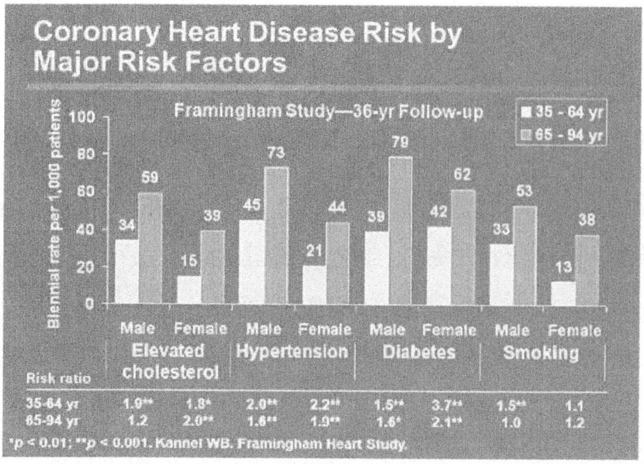

Chapter 7
Cardiovascular Drugs

Superstition vs. Medical Science

When pharmacotherapy emerged in ancient times, it was often rooted in the belief that sickness was caused by an angry deity and that appeasing the deity was essential for a cure. In the Babylo-Assyria era, water and incantations to deities were the major elements of *materia medica,* and the use of botanicals began. The ancient Egyptians used herbal products in an expanded form of polypharmacy, the use of multiple drugs for treatment. The pagan polytheism of the ancient Greeks/Romans included invocation to their key gods — the modern physician's Rx symbol is, in fact, derived from the symbol for the (Greco-) Roman king of the gods, Jupiter, whose name was invoked before prescribing medications.

Over the centuries, scientific progress helped to drive superstitious fatalism out of medicine.

The 20th Century and Cardiovascular Drugs

In the 20th century, especially its second half, previously untreatable diseases became treatable — and some even became curable. Marked changes occurred in the preparation and distribution of cardiac and other remedies. The most striking shift was from the physician-pharmacist working with mortar and pestle in an apothecary to scientists working with sophisticated equipment in huge pharmaceutical establishments. In the U.S., there was the growing collaboration of the pharmaceutical (pharma) industry with academia. The greater part of the academic budget was devoted to basic research, i.e., cellular and molecular science, and industry was encouraged to share the financial burden of clinical research, as in the trial of new drugs and devices — especially as pharma companies stood to benefit from those that proved to be safe and effective.

The second event that enhanced collaboration in the U.S. was an amendment to the patent law in 1980 that allowed the academic institutions to hold patent rights. Academia could now patent an invention and by eliminating government control, deal directly with industry to commercialize it — this put it at a par with international competitors. The new industry of biotechnology would emerge due to this change. More importantly, a truly scientific atmosphere came into being for the development of physiologically effective drugs. Research and development (R&D) is the hallmark of the pharmaceutical industry, a fiercely competitive field — undoubtedly, this competitive R&D has been of great benefit to people.

The 20th century saw an avalanche of medications developed for the treatment of heart disease — beta blockers for treatment of angina, or heart rhythm abnormalities; ACE inhibitors for high blood pressure; statins for lipid abnormalities; and clot busters, i.e., fibrinolytic agents, for clot dissolving in coronary artery disease. Let us now discuss some of the pioneers and their great efforts and accomplishments.

Beta Blockers

(Sir) James Black (1924-2010), a Scottish physician and pharmacologist, made key contributions to the treatment of peptic ulcer, coronary heart disease and more. He was inspired in his pharmacology research work by his mentor, **D'Arcy Wentworth Thompson**, author of a classic 1917 study, *"On Growth and Form,"* and an expert on variations in the chemical behavior of crystals of a similar shape. Black conducted research on various substances' effects on specific cell receptors — and the relationship to blood flow and intestinal absorption. This led to his understanding the fundamental mechanism that underlay his important discoveries, which we shall now discuss.

Black knew that to test patients for susceptibility to peptic ulcer, physicians would administer histamine, a powerful stimulant for gastric secretion. He used this knowledge to develop a drug in the 1960s that would prevent injurious secretions of acid: The drug, cimetidine, radically changed peptic ulcer treatment.

Equally revolutionary around that same time was the development of beta blockers: Black accomplished this while working with the British pharmaceutical company, Imperial Chemical Industries. Beta blockers could be used clinically without any significant adverse effects.

The first such compound was propranolol, marketed under the trade name Inderal. Propranolol's molecules would interact with proteins in cell receptors on the surface of or within the cell to block the cell's behavior.

(L to R) James Black, David Cushman/Miguel Ondetti and Akira Endo

Hailed as the greatest breakthrough in heart disease treatment since the discovery of digitalis (digoxin, a drug used to treat various heart conditions), propranolol/Inderal benefited patients with coronary artery disease, arrhythmias, hypertension and heart failure.

In 1988, Black's work was honored, along with that of researchers **George Hitchings** and **Gertrude Elion**, with the Nobel Prize in Physiology or Medicine, for *"discoveries of important principles for drug treatment."*

ACE Inhibitors

Angiotensin-converting enzyme (ACE) inhibitors have become a cornerstone in the treatment of hypertension, or high blood pressure, and of heart failure. Uncontrolled hypertension is a risk factor for the development of coronary artery disease and heart attack. Also, it adversely impacts the heart function both in the contraction (systolic) and relaxation (diastolic) phases of the heartbeat. As such, hypertension is an important cause of impairment of heart function, leading to heart failure.

MIGUEL ONDETTI and DAVID CUSHMAN
Synthesized the First ACE Inhibitor, Captopril

Miguel Ondetti and David Cushman's collaborative work at the Squibb Institute for Medical Research in New Jersey resulted in synthesis of the first ACE inhibitor, captopril, an anti-hypertensive drug.

A native of Buenos Aires, Argentina, Ondetti as a young man won a scholarship to work on his doctoral dissertation at Squibb's facility there. * After earning his PhD from the University of Buenos Aires, Ondetti in 1960 joined Squibb in New Jersey. There, he worked in peptide synthesis research. Cushman joined Squibb in the late 1960s, after receiving his PhD in biochemistry from the University of Illinois.

Some years later, Cushman and Ondetti met Nobel Prize winner John Vane and were influenced by his work. His team at the Royal College of Surgeons in London had shown that peptides from the Brazilian pit viper's venom inhibited the activity of ACE in animal lungs — by blocking conversion of angiotensin 1 to angiotensin 11 and by activating the peptide bradykinin to do its work of dilating blood vessels. So, the viper's bite caused death from the resulting drop in blood pressure. The researchers correctly theorized that the same peptides might be useful in treating hypertension. ◆

** In exchange for the privilege of monopoly rights to manufacture antibiotics in Argentina, Squibb was required to invest some of its profits in the country and thus, in the institute and scholarship.*

In the 1970s, chemist **Miguel Ondetti (1930-2004)** and his biochemist colleague **David Cushman (1939-2000)** at the pharmaceutical company R.R. Squibb (now Bristol-Myers Squibb) together developed a new means of treating hypertension. They did so using the principles of rational drug design pioneered by Black, Hitchings and Elion. Ondetti and Cushman found a specific ACE inhibitor and eventually used it to produce captopril, the parent drug to what would be a slew of newer drugs in cardiovascular medicine. In 1999, Ondetti and Cushman received the Albert Lasker Clinical Medical Research Award for their contributions in this area. They were also honored by the American Chemical Society.

Clot Busters – Fibrinolytic Agents

An exciting new development was the introduction in **1960** of clot busters or clot dissolving agents, i.e., fibrinolytic agents for thrombolytic therapy, for coronary artery disease. **Sol Sherry**'s group reported a rational approach to thrombolysis using intravenous streptokinase.

It was a commonly held view that progressively diminished blood flow due to atherosclerosis, or 'hardening' of the coronary arteries, led to death of the heart muscle and that blood clots in these arteries developed after a heart attack. William C. Roberts, chief of cardiac pathology at the National Institutes of Health, held, *"Although it may play a major role in causing atherosclerosis, coronary thrombosis may well play a minor role or none at all in precipitating a fatal coronary event."*

However, **Marcus DeWood** made a remarkable discovery by providing angiographic evidence that total occlusion by blood clots of infarct-related coronaries in the early period of myocardial infarction was the cause of a heart attack. **Yevgeniy (Evgeny) Chazov** in the Soviet Union, mentioned earlier, and **Peter Rentrop** in West Germany separately demonstrated the rapid recanalization and restoration of blood flow after local administration of streptokinase directly into an infarct-related blocked artery.

Through recombinant techniques where different strands of DNA are combined to produce a designer molecule, a new drug called tPA (tissue plasminogen activator) was similarly shown to open many of the blood clot-blocked coronary arteries. Both tPA and streptokinase improved a patient's likelihood of surviving a heart attack, a major advance enthusiastically embraced by the medical world. The advent of angioplasty, to be described shortly, was the next major advance in the treatment of heart attack.

Cholesterol and Statins (Cholesterol-Lowering Drugs)

Cholesterol, a kind of lipid (fat) made by all animal cells, most of all in the liver, is essential for cellular structure and the functioning of human organs. (It is also necessary for the body to make steroid hormones, bile acids and vitamin D.) However, during a century of investigation, scientists have established a causal connection among blood cholesterol, atherosclerosis and coronary heart disease.

Building on that knowledge, scientists and the pharmaceutical industry have developed an effective class of drugs — the statins — to block the action of a liver enzyme that helps produce cholesterol. This blocking can lower cholesterol levels in the blood and, in many people, reduce the frequency of heart attack.

Since it was first isolated from gallstones in 1784, cholesterol has fascinated scientists in many areas of science and medicine. In fact, 13 Nobel Prizes have been awarded to scientists who devoted major parts of their careers to cholesterol research. Here is some of the history behind these advances.

In the 1950s, studies linked serum cholesterol to heart disease. In 1956, researchers isolated mevalonic acid, a key precursor of serum cholesterol. Three years later, scientists at the Max Planck Institute/Society in Germany discovered the enzyme HMG-CoA reductase, which regulated the key step in mevalonic acid production. The first HMG-CoA reductase inhibitor, a statin, was isolated by biochemist **Akira Endo** in Tokyo, Japan: Called compactin, the drug was soon found to lower cholesterol levels in experimental animals.

Endo was born in 1933, raised on a farm in northern Japan and from his youth, intrigued by fungi. He admired Alexander Fleming's famous work on fungi that had ultimately led to the discovery of penicillin as a life-saving antibiotic. Early in his career, Endo conducted applied microbiology research on fungi for a pharmaceutical company in Japan and related to this, became interested in the biosynthesis of cholesterol —

certain mushrooms and other fungi naturally produce a type of statin. On a leave of absence from his job in Japan, Endo worked with **Bernard Horecker** and **Lawrence Rothfield** at the Albert Einstein College of Medicine in New York. While there, he learned about high cholesterol as a major risk factor for the development of cardiovascular disease.

Endo returned to Japan and continued his work on cholesterol — and his interest in and work with fungi and cholesterol ultimately led, in the 1970s, to the discovery of statins. Endo's findings were honored with major awards, including the Japan Prize in 2006 and the Lasker-DeBakey Clinical Medical Research Award in 2008. He was presented the Canada Gairdner International Award in 2017 for: *"The first discovery and development of statins, inhibitors of cholesterol biosynthesis that have transformed the prevention and treatment of cardiovascular disease."*

Michael Brown (l) and Joseph Goldstein in their lab

Over three decades, **Michael Brown** and **Joseph Goldstein** at the University of Texas Health Science Center and Southwestern Medical School in Dallas, studied the low-density lipoprotein (LDL) cholesterol pathway. This work was part of their research on identifying the genetic defect in familial hypercholesterolemia.

In 1985, they were jointly awarded the Nobel Prize in Physiology or *Medicine "for their discoveries concerning the [genetic] regulation of cholesterol metabolism.*
Brown/Goldstein and Endo have generously credited each other for influencing their research and findings.

Meanwhile, scientists at the pharmaceutical company Merck Inc. discovered an agent that was a pure inhibitor of HMA CoA reductase, and they named it lovastatin. In 1991, pravastatin (Pravachol), a derivative of compactin, and simvastatin (Zocor), a synthetic derivative of lovastatin, were both approved by the U.S. Food and Drug Administration (FDA).

Collectively, the statins have become some of the most widely prescribed medications in the world. In 1994 the *Scandinavian Simvastatin Survival Study* showed that among a group of patients treated with simvastatin, 30 percent or greater experienced a reduction in mortality, coronary events or the need for angioplasty or bypass surgery.

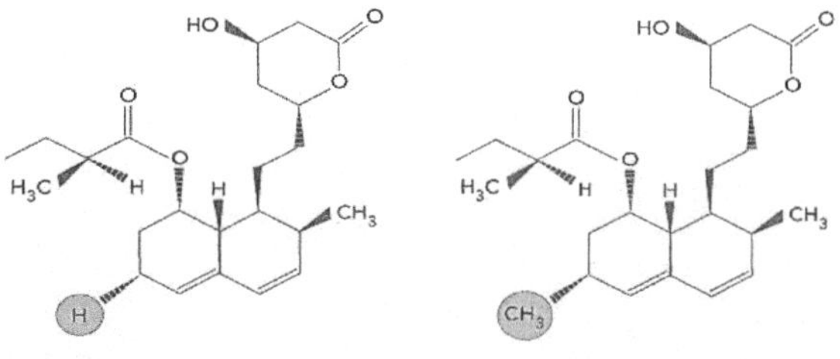

(L to R) Compactin, the first statin, discovered by Akira Endo; Lovastatin, first statin approved for clinical use, developed by Merck

Subsequent studies conducted using lovastatin confirmed the results for patients both with and without a prior history of coronary artery disease, as well as those with a history of diabetes, peripheral vascular disease and stroke.

Beta blockers, ACE inhibitors and statins have improved and prolonged the lives of hundreds of millions of patients throughout the world. From the 1990s onward, with the powerful tools of genetic engineering, there has been a greater promise of additional and even more efficient drugs.

Chapter 8
Diagnostic Ultrasound: Echocardiography

Visualization of the heart and great vessels by noninvasive imaging now makes many invasive procedures unnecessary. First came echocardiography, the development of which is described in the following pages. Subsequently came a variety of nuclear techniques as well as advanced radiological techniques, i.e., CT (computerized tomography) scans and MRI (magnetic resonance imaging).

By permitting sequential testing, such imaging allows optimal timing of interventions and assessment of treatment response. Noninvasive imaging represents an enormous advance in diagnosing heart disease and caring for patients.

Echocardiography uses ultrasound waves to create an image of the heart and the pattern of blood flow(ing) through it. Ultrasound is a frequency of vibrations above the range audible to the human ear. The resulting echocardiogram is useful for measuring the heart's size and strength and for evaluating the heart valves.

The concept behind echocardiography was first demonstrated by **Lazzaro Spallanzani** in the 1700s. He demonstrated that bats were able to navigate by using reflected echoes of inaudible sounds. The ability to create ultrasonic waves was made possible in 1880 with the discovery of piezoelectricity (electrical charges induced in a crystalline substance by the application of pressure) by the physicists, and brothers, **Jacques Curie** and **Pierre Curie**.

An echo ranging technique to detect underwater objects resulted in the development of SONAR (*So*und *N*avigation *a*nd *R*anging) and was suggested by **Lewis Richardson** and later developed by physicist **Paul Langevin** in 1915. (The acoustic frequencies used in SONAR systems vary from very low, or infrasonic, to extremely high, or ultrasonic.) SONAR development occurred in time for military forces to detect enemy submarines during WWI. By the time of WWII, the U.S. Navy was using reflected radio waves to detect the presence of enemy airplanes, a technology termed RADAR (*Ra*dio *D*etection *a*nd *R*anging). The concept therefore emerged that ultrasound might also be used to visualize internal organs of the body, as in the case of the heart, is described later in the book.

Heart surgery for mitral stenosis — a narrowed heart valve, causing increasing shortness of breath, palpitation and, when advanced enough, death — was initiated in the late 1940s. The surgery involved dilating the constricted valve by inserting a finger through it, in a procedure called a 'finger fracture,' closed mitral commissurotomy. The trouble was that these same symptoms of mitral stenosis could also be caused by a defective mitral valve, with mitral regurgitation — which was worsened by surgery.

The correct identification of pure or predominant mitral stenosis and its distinction from mitral incompetence was therefore critical when considering heart surgery. The only reliable way to do this was by cardiac catheterization, which at the time was an invasive and elaborate technique offering inadequate imaging techniques. In the 1950s, cardiac catheterization was still in its infancy and failed to provide an accurate appraisal of the status of the mitral valve.

Echocardiogram

Inge Edler 1911-2001), director of the cardiac catheterization laboratory at Lund University Hospital in Lund, Sweden, was challenged by these difficulties and the need to provide a more accurate assessment of defects of the heart using simpler and noninvasive methods. Edler thought of exploring ultrasound for this purpose, and he pursued it in collaboration with **Carl Hellmuth Hertz (1920-90)**. Thus, echocardiography was born. (The actual term "echocardiography" was most likely first used in the mid-1960s by cardiologist **Bernard Segal** and then became widely accepted.)

In 1953, using a Siemens ultrasonic reflectoscope, Edler and Hertz recorded the first moving pictures of the heart, thus inaugurating the field of diagnostic ultrasound for the heart, or echocardiography. Their article titled, "The use of ultrasonic reflectoscope for continuous recording of the movements of the heart valves" was published in *The Proceedings of the Royal Physiologic Society* in Lund, Sweden.

Inge Edler (r) with Carl Hellmuth Hertz and the Siemens ultrasonic reflectoscope they used in their research

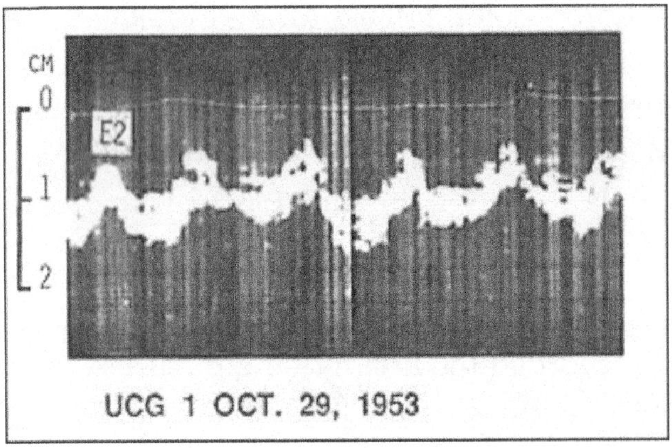

Picture showing heart valves moving — first-ever such image; captured by Edler and Hertz using an ultrasonic reflectoscope

To confirm his observations, Edler subsequently conducted ultrasound studies on dying patients. On completion of his examination, he would mark the direction of the ultrasonic beam on the patient's chest. After the death of the patient, he would pass an ice pick through the chest wall in the direction of the previously marked ultrasonic beam. The needle would pierce the anterior wall of the right ventricle, pass through the right ventricular outflow tract, the interventricular septum, the upper part of the left ventricle, the anterior leaflet of the mitral valve and the left atrium. He thus realized that the echo signals coming from the anterior leaflet of the mitral valve in mitral stenosis were different from those in mitral regurgitation.

Edler later used the technique primarily for the preoperative study and diagnosis of mitral stenosis and of mitral regurgitation. He was also able to diagnose pericardial effusion, an accumulation of fluid around the heart. In addition, he continued his studies on mitral valve disease. Edler's work was demonstrated in a scientific film shown at the Third European Congress of Cardiology in Rome in 1960 and published in 1961.

In collaboration with cardiologist **Niles-Rune Lindstrom**, Edler introduced the use of combined Doppler ultrasound and echocardiography at the First World Congress on Ultrasonic Diagnostics in Medicine in Vienna, especially useful in the diagnosis of aortic and mitral valve regurgitation. Hertz created new techniques for two-dimensional echocardiography and for using the Doppler effect to measure the direction and velocity of the flow of blood and cardiac tissue.

Doppler echocardiography is a type of echocardiography based on a physics concept first described in 1942 by Austrian mathematician and physicist Christian Doppler (1803-53): the Doppler effect, which was named after him. The Doppler effect is a change or shift in frequency in relation to the movement of the source. Doppler the physicist demonstrated that the frequency of sound reflected from an object is altered as the object is moving. For example, a person might hear a high-pitched note if the source (as in a whistling train engine) were moving toward him and a low-pitched note if it were moving away from him. Doppler echocardiography capitalizes on the fact that blood cell movements produce the Doppler shift of ultrasound frequencies — the extent and direction of shift are related to the velocity and direction of blood flow.

INGE EDLER and CARL HELLMUTH HERTZ

Pioneered the Use of Ultrasound for Diagnosing Heart Abnormalities – the Echocardiogram

Inge Edler and Carl Hellmuth Hertz are known as pioneers in clinical ultrasonography and echocardiography. How their fruitful collaboration came about, and proceeded, is also quite interesting.

In 1953, Edler was head of cardiology at Lund University Hospital in Lund, Sweden, and responsible for preoperative diagnosis of heart disease. At that time, cardiac catheterization and contrast x-rays of the heart failed to give enough data for a correct appraisal of the mitral valve's status. Seeking a non-invasive alternative, Edler wondered if high-frequency sounds could be generated for short distances. (He knew about echolocation and the use of sound and radio waves to locate objects, including WWII era use of sonar to detect enemy submarines and radar to detect enemy airplanes.)

Edler was introduced to Hertz, who was then a graduate student and junior assistant in nuclear physics at the university with a special interest in ultrasound, or high-frequency sound. This introduction led to what would become a close collaboration between the two, in search of a way of 'visualizing the heart.' Soon after meeting, Edler and Hertz began their work. First, they sought access to an ultrasonic reflectoscope. Hertz was familiar with one developed for testing nondestructive materials at a shipbuilding yard in Malmo. In 1953, he visited the company and was able to borrow their reflectoscope. Back at the university's cardiac laboratory, Edler and Hertz began experimenting with the reflectoscope.

They put the ultrasound sensor, or probe, over Hertz's own heart – and to their great excitement, they saw well-defined echoes moving back and forth along the x-axis of the oscilloscope screen, at a depth of 8-9 cm. from the chest wall, in synchrony with is heartbeat. The machine recorded these echoes from Hertz's heart. Later, through a personal connection, Hertz obtained an advanced ultrasonic reflectoscope on loan from Siemens Laboratories in Germany. He and Edler were thus able to advance their research work in echocardiography. *

Edler's contributions to medicine received world recognition and honors. In 1977, he and Hertz together received the Lasker-DeBakey Clinical Medical Research Award, *"For pioneering the clinical application of ultrasound as a noninvasive tool in the medical diagnosis of abnormalities of the heart, probably the most important noninvasive tool for cardiac diagnosis since the electrocardiography machine."* ◆

** Carl Hellmuth Hertz was the son of Gustav Ludwig Hertz, a Nobel laureate in physics who during WWII had served as director of Siemens Laboratories. Gustav's uncle Heinrich Rudolf Hertz, also a physicist, proved the existence of radio waves as electromagnetic radiation waves. The unit of frequency called "hertz" (Hz), or cycle(s)/sec., was named for Heinrich Hertz.*

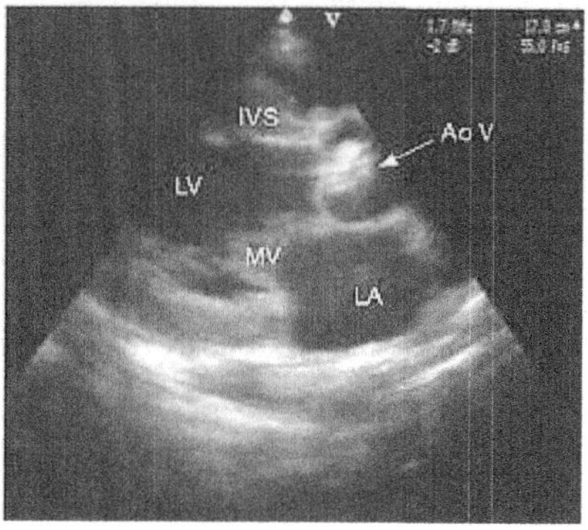

Echocardiogram

For some time, no further advances were made in this area. Then came **Harvey Feigenbaum**'s pioneering work on developing modern echocardiography. Feigenbaum sought answers in ultrasound, as he was frustrated with the tediousness and inaccuracies of using catheterization techniques to measure cardiac output, volume and pressures. His work attracted many young cardiologists to his basement laboratory at the Krannert Institute in Indianapolis, Indiana, where he and associates developed new applications. A number of these cardiologists, with their newly acquired knowledge, then set up their own investigative centers and training programs.

Doppler echocardiography has become an important component of complete echocardiographic evaluation. Furthermore, an adequate graphic spectral display has become an important addition to echocardiography. It has helped in delineating abnormal blood flow within the heart, as in defective heart valves or in congenital heart disease — a "hole in the heart," for example, in atrial or ventricular septal defects. It has also helped in assessment of heart valves, leaking through valves, evaluation of cardiac output and assessment of the heart's (ventricular) function.

Noninvasive imaging to visualize the heart and great vessels, like that initiated by echocardiography, represents an enormous advance in diagnosing heart disease and caring for patients. It has made many invasive procedures unnecessary and enhanced the timing of interventions and assessment of treatment response.

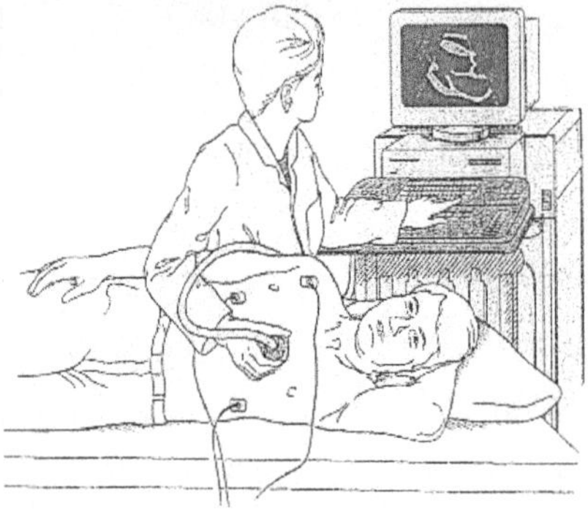

An echocardiography procedure

Chapter 9
Interventional Cardiology

Interventional cardiology comprises procedures that reduce or eliminate blockages in coronary arteries so that blood flow (with oxygen and other nutrients) can be restored to blood-deprived heart tissue. This then lessens chest pain, diminishes damage to the heart muscle and reduces the need for heart medications, in most patients.

These catheter-based procedures, such as angioplasty and stenting, successfully prop open blocked coronary arteries. The procedures have improved and extended many people's lives.

In the early **1960s,** vascular radiologists **Charles Dotter** and **Melvin P. Judkins** at the Oregon Health Sciences University in the U.S. were the first to recognize that the lumen of (narrowed) peripheral blood vessels could be enlarged by passing a guidewire and then a catheter or rigid dilator over the narrowed area. They described this technique — which came to be known as angioplasty — for relieving stenosis of iliofemoral (leg) arteries. The difficulty with the Dotter technique was the trauma caused by introducing rather large caliber rigid dilators. Regardless, Dotter is known today as one of the fathers of interventional and vascular radiology.

Percutaneous Transluminal Coronary Angioplasty

German-born radiologist-cardiologist **Andreas Gruentzig (1939-85)** learned of Dotter's angioplasty work, while at a lecture in Frankfurt. Gruentzig wanted to explore the angioplasty technique but encountered bureaucratic resistance in his native land, so he moved to Zurich, Switzerland. There, he worked in his home kitchen-based lab on developing balloon catheters. The breakthrough came in 1974 when Gruentzig and **David Kumpe** — while performing a percutaneous transluminal coronary angioplasty (PTCA) procedure — substituted a balloon-tipped catheter for the rigid dilator initially used for obstructive lesions involving the peripheral arteries.

Gruentzig described the balloon technique as one in which the obstruction remains flattened against the walls of the blood vessel, like "footprints in the snow." A catheter tipped by a tiny deflated balloon is introduced through locally anesthetized skin (percutaneous) into a peripheral artery (transluminal) and advanced to the narrowed or obstructed coronary artery. The balloon is inserted, and when inflated, pushes the fatty deposits or clot against the side of the artery.

Andreas Gruentzig's laboratory in his home kitchen

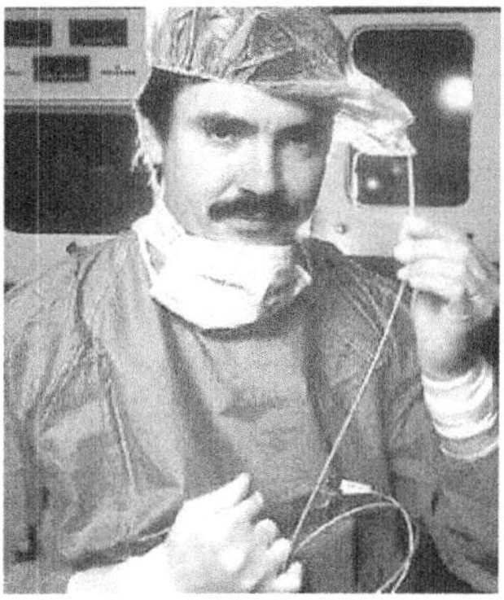
Andreas Gruentzig, with his balloon-tipped catheter

This forces an expansion of the inner white blood cells/clot plaque deposits and the surrounding muscular wall, opening the blood vessel for improved flow (angioplasty). The balloon is then deflated and withdrawn. The procedure enables the patient to avoid bypass surgery and allows a quick recovery.

In 1977 Gruentzig performed PTCA on an awake human, a 38-year-old businessman named Adolf Bachmann (of the same age as Gruentzig). The patient was a two-pack a day cigarette smoker with class 4 angina, and he had refused coronary artery bypass surgery. Two months later, angiograms of the PTCA procedure were presented at an American Heart Association meeting — to a stunned audience of cardiologists, who burst into applause, followed by a standing ovation. The patient quit smoking, discontinued cardiac medications and years later, began taking aspirin and cholesterol-lowering drugs. His angiogram in 1987 showed continued excellent results.

After 23 years of a pain-free and enjoyable life, Bachmann felt chest pain due to a blockage at the original site. He underwent successful angioplasty and stent placement. In 2007, many years after these procedures, Bachmann was invited to a Transcatheter Cardiovascular Therapeutics Conference held in Washington, DC.

There, he spoke to a large audience in a ballroom and needed no introduction. *"I feel compelled to give you all a big hug,"* he said to the group. *"Today I am active, singing in two choirs, hiking, gardening and playing cards and pain free, while only taking aspirin."*

Gruentzig had expanded the use of the cardiac catheter, until then used as a diagnostic tool, and made it a powerful therapeutic device. His innovative technique took the cardiology world by storm. In subsequent years, he performed coronary angioplasty in 169 patients, while conducting a training program for cardiologists from across the globe. In no time at all, PTCA became a method of choice for treating coronary artery obstruction. In addition to treating coronary stenosis, angioplasty balloon catheters can be used to open almost any abnormal obstruction in the heart and circulation. Furthermore, many abnormal openings can be closed using catheter-based techniques.

Balloon angioplasty was later followed by stenting with bare metal stents, which have now been replaced by drug-eluting stents. Within a few short years, interventional cardiology was established as a subspecialty of cardiology. Following Gruentzig's introduction of coronary angioplasty for single vessel disease, **Geoff Hartzler** extended its use to multivessel disease and, importantly, to use in acute myocardial infarction. He thereby revolutionized the treatment of heart attack.

But it was Gruentzig who began the revolution in treatment of coronary artery disease, opening a new world of successes — and challenges. Gruentzig's masterpiece move with the first coronary angioplasty was, of course, built on the shoulders of his predecessors, including Werner Forssmann, Charles Dotter, Mason Sones and others. Related concepts have continued to become reality.

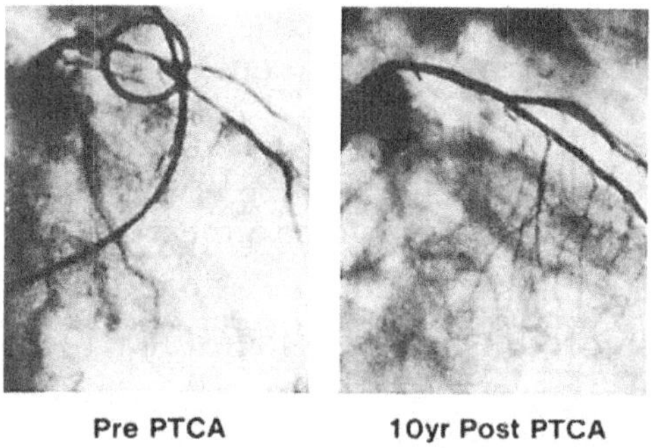

Treating coronary artery disease with PTCA:

(L to R) Circled area at top left shows narrowing of artery; next image shows blood flow restored and artery unobstructed even 10 years after PTCA.

ANDREAS GRUENTZIG

Put Interventional Cardiology on the Map: Conducted First Successful Coronary Angioplasty

Working in Switzerland, Andreas Gruentzig devised a special type of balloon catheter, and in 1977, used it in the first successful coronary angioplasty (PTCA). His balloon catheter enlarged narrowed arteries without the need for surgery.

The balloon catheter had two lumens: One lumen was devoted to monitoring the injection of radiopaque material for visualizing the coronary tree; the other lumen was used as a means of inflating or deflating the balloon. A small flexible guide wire attached to the tip of the catheter served as a screening mechanism. Gruentzig's catheter could be guided into the stenotic area with the balloon in a deflated state and when in place, inflated to a predetermined size. His modification to earlier catheters was pivotal and gained him a well-deserved place among the leaders in cardiology.

Gruentzig was an interesting man: His qualities of collaborativeness and trustworthiness were, reportedly, coupled with a deceptive ambition: He worked collaboratively to persuade surgeons to use his invention. Although it was a challenging task — surgeons were to be most directly affected, and surgeons and cardiologists were potential competitors — yet, he eventually succeeded.

Gruentzig continued to improve his technique and treat patients while teaching cardiologists from around the world. Three cardiologists from the Emory University School of Medicine in Atlanta, Georgia who attended Gruentzig's course in Switzerland learned of his desire to relocate to the U.S. This conversation resulted in a successful effort by Emory to bring Gruentzig on board: He became a faculty member in medicine and radiology and director of Interventional Cardiovascular Medicine.

As a doctor, Gruentzig was deeply caring of his patients. He also loved to teach and was adored by his students. In addition, he sang, danced and charmed many. He married, got divorced and then remarried a medical student. Gruentzig bought a spectacular mansion where he held lavish parties. He also bought his own airplane and a cottage on Sea Island, Georgia. In 1985, as he was flying his plane back to Atlanta during a major storm, the plane crashed: Tragically, the doctor, 46, and his wife were killed.

Emory established the Andreas Gruentzig Cardiovascular Center to honor his contributions to interventional cardiology. Cardiologists working in the Gruentzig Center have made many contributions in this rapidly expanding field, including, in 1987, the first use in the U.S. of a coronary stent to assure the patency (a state of continued expansion) of a clogged coronary artery opened up by balloon angioplasty. ◆

Chapter 10
Pacemakers and Defibrillators

The human heart usually beats at around 50-70 times per minute at rest. In some heart conditions, the heart may slow down significantly (bradycardia), and if the heart rate falls below 40 per minute, the brain and body may suffer from hypoxemia, a low oxygen level from insufficient blood supply. This usually results in symptoms such as dizzy spells, fatigue or shortness of breath. It may also result in cardiac arrest, a sudden cessation of heart's functional circulation or even stoppage of heartbeat.

Contraction of heart muscle (heartbeat) is initiated by rhythmic electrical impulses generated by heart muscle cells called pacemaker cells, which directly control heart rate. When the process goes awry, due to disease affecting the pacemaker cells and/or the electrical conduction pathways, external assistance is needed. The artificial pacemaker of today is a small device placed in the patient's chest or abdomen to help control heart rhythm. This device uses electrical impulses to prompt the heart to beat at normal rates. Its development resulted from important work done in the mid-20th century onward.

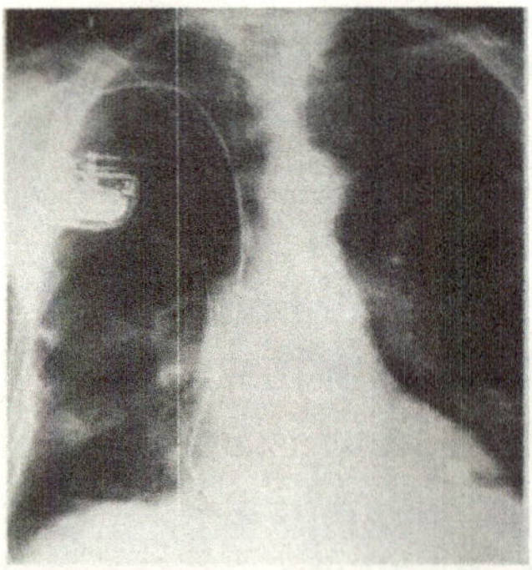

Chest x-ray showing pacemaker in place in heart's right ventricle

The impetus for stimulating the heart by artificial means — cardiac electrostimulation — was triggered by advancement of heart surgery in the mid-20th century: As a complication of the surgery, heart block could result and cause a slowing of the heart.

In the **1950s**, continuing the work of the electrophysiologists preceding him, **Paul Zoll (1911-99)** in Boston built the first external cardiac pacemaker. Located outside the body, the device could trigger mechanical contraction of the heart by emitting electronic discharges transmitted through the chest wall. His important paper published in 1952 described this external method of cardiac electrostimulation, based on his demonstrating that the heart could be electrically stimulated to beat effectively.

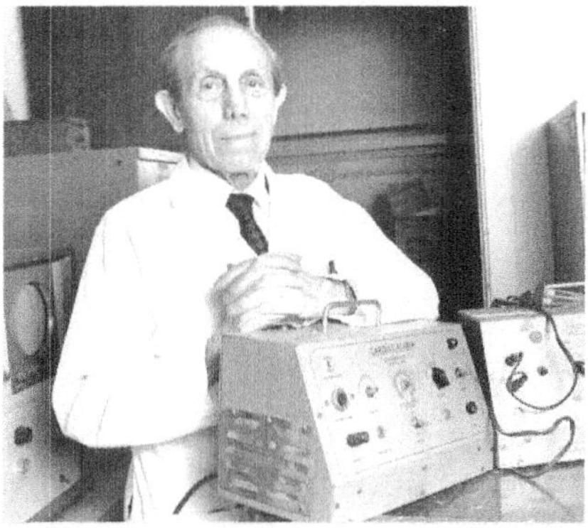

Paul Zoll with his external cardiac pacemaker

Zoll's advance eventually helped in development of implantable cardiac pacemakers, as well being an asset in the coronary care unit. In 1973, he received the Albert Lasker Clinical Medical Research Award in recognition of his key contributions.

Implantable/Internal Cardiac Pacemaker

Rune Elmqvist (1906-96) and **Åke Senning (1915-2000)** in Sweden, in 1958, reported the first use of an internal cardiac (epicardial) pacemaker, which was implanted after a thoracotomy procedure, a surgical incision into the chest wall. *(See story on following page).* The first model had to be replaced after only a few hours. The technology did rapidly improve, however, and models of increasing complexity met specific requirements of a malfunctioning heart.

Today, internal pacemakers with a lifespan of 10 years or longer are standard in cardiology. Modern technology has transformed what began as external devices into totally implantable devices with miniaturization of electronic circuitry. The size of a wrist watch, the artificial pacemaker consists of: a generator unit (battery), computer circuits capable of monitoring the patient's heart rate and delivering an electrical impulse leading the heart to contract, or beat, at a desired rate, and wires that connect the pacemaker to the heart.

Pacemakers implanted in patients with slow heart rhythms can restore life expectancy to normal levels while retaining the ability for normal activities. Prognosis is determined by whether they have coexisting heart disease. An average battery lasts 5 – 15 years, patients periodically check in at a cardiologist's office, and in some cases, telephone monitoring ensures timely battery replacement. Currently about 600,000 pacemakers are implanted annually, and more than 3 million people worldwide live with pacemakers.

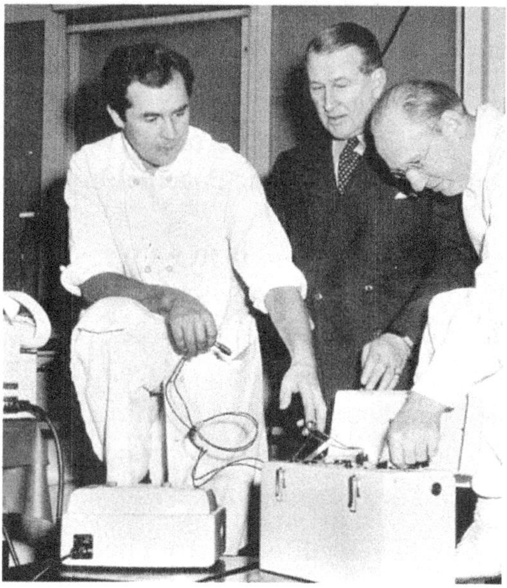

Conducting early research on cardiac pacemakers: (L to R) Åke Senning, Rune Elmqvist and Clarence Crafoord

Implantable cardiac pacemaker developed by Elmqvist and Senning

RUNE ELMQVIST and ÅKE SENNING

Pioneered the Use of an Implantable Cardiac Pacemaker

In 1958, the wife of a Swedish man with severe cardiac arrhythmia read in a newspaper that a team at Stockholm's Karolinska University Hospital was working on an implantable cardiac pacemaker. The team included heart surgeon Åke Senning, his mentor and fellow heart surgeon Clarence Crafoord, and an MD turned engineer, Rune Elmqvist.

Arne Larsson had been suffering from heart arrhythmia due to a viral infection. His heart was beating at 28 beats per minute (a healthy heart beats at around 70 beats per minute). He was frequently losing consciousness and had to be revived 20-30 times a day. His wife Else-Marie refused to accept her husband's impending fate.

Up to then, the team at Karolinska had experimented only on animals. But Else-Marie approached the team and pleaded with them to save her husband's life. So, the surgical team proceeded: Arne received his first pacemaker in a secret and successful emergency operation. He lived to be 86 years old, outliving the team members who had saved his life. During his lifetime, he received 20 pacemaker replacements. Thanks to these small devices, he was able to resume an active life that included cycling, dancing, sailing and more. ◆

Defibrillators

As noted earlier, **Paul Zoll** and his associates performed the first successful external defibrillation in a human subject in **1956.** Subsequently, **Michel Mirowski (1924-90)** conceived of the automatic implantable cardioverter defibrillator (ICD), after his mentor died of a recurrent ventricular tachycardia, a dangerous heart arrhythmia. Mirowski invented the ICD while working at Sinai Hospital in Baltimore, Maryland. In 1980 in the *New England Journal of Medicine,* he reported its first successful human application.

Michel Mirowski

The FDA cleared the device for commercial use in the U.S. in 1985 — about 19 years after Mirowski conceived of the battery-operated device. It was the first successful therapy for out-of-hospital cardiac arrest. Now miniaturized and much improved, the ICD device is considered an important tool in preventing sudden cardiac death both in primary and secondary heart disease.

MICHEL MIROWSKI

Invented the Automatic Implantable Cardioverter Defibrillator

Michel Mirowski was truly a man of the world: Born in 1924 in Warsaw, Poland, he attended medical school in France, completed his residency in Israel, and did his cardiology fellowship at Johns Hopkins Hospital in Baltimore, Maryland and the Institute of Cardiology in Mexico City. Eventually, he became the first medical director of the Sinai Hospital in Baltimore.

Working long and hard, Mirowski invented the automatic implantable cardioverter defibrillator (ICD). In 1980, a decade later, Mirowski reported successful application of the ICD.

The first ICD patient was a 57-year-old woman with documented coronary artery disease, a history of acute myocardial infarction, recurrent syncope with documented ventricular fibrillation. She was treated at Johns Hopkins Hospital, with Levi Watkins as the operating surgeon, working with Mirowski's team. The patient was given general anesthesia, and the procedure was going well — until the ICD was requested. The circulating nurse picked up the package containing the device but then dropped it on the floor, where it broke. Fortunately, a second device was available, and it was successfully implanted. ◆

II. 21st Century: Today, and Looking Forward

i. The Present Day

Cardiology advanced tremendously between the 20th and the 21st centuries, as the preceding descriptions have shown. The medical, surgical and electrophysiological achievements of the 20th century have provided great benefits — unimaginable some decades ago — to millions around the world.

One public figure, for example, who has benefited immensely from these advancements is former U.S. Vice President Dick Cheney. He has been treated primarily at the George Washington University Hospital in Washington, DC, by cardiologist Jonathan Reiner, ably assisted by a team of physicians and nursing staff. His later care has also included additional treatment received at the Inova Fairfax Hospital. Cheney suffered his first heart attack in 1978, as a relatively young man, and subsequently benefited from nearly every medical breakthrough. He lives on today, 4 decades later; as of the writing of this book, he is 79 years old. Cheney has been presented as a case study on how innovations in medical science and technology have greatly extended human lives.

Electricity and the Heart

Some major advances have occurred in the arena of electrophysiology. A heart ailment increasingly being recognized, especially in older patients, is atrial fibrillation, an irregular heartbeat that increases the risk of stroke and heart-related complications. Atrial fibrillation is the most common arrhythmia in the elderly, affecting about one in 10 individuals over the age of 65. Treatment includes lifestyle changes, medications, cardioversion and more recently, procedures such as ablation, pacemakers or surgery. Catheter ablation, where a catheter threaded into the heart to apply heat or extreme cold causes tiny scars in the heart muscle to disrupt or eliminate erratic electrical impulses, has proved to be effective.

For those with non-valvular atrial fibrillation, the Watchman procedure may be considered in carefully selected patients with prohibitive bleeding risks with anticoagulation. It involves a permanent heart implant into the left atrial appendage.

Cardiac pacemakers and implantable cardiac defibrillators with increasingly sophisticated advances have become important therapeutic modalities for patients with life-threatening ventricular arrhythmias and dysfunction. A leadless pacemaker implanted with a trans-catheter pacing system with an estimated battery life of 12 years is currently helping patients who need pacing in only one chamber and has a promising future. It may advance to dual chamber and then to three-chamber resynchronizing devices in left ventricular pacing.

Cardiovascular Drugs

Prime examples of progress in cardiovascular drugs or medications are: ACE inhibitors used for the control of high blood pressure, statins in the treatment of elevated cholesterol, and "clot busters" or fibrinolytic and thrombolytic agents.

With the introduction of non-vitamin K antagonist oral anticoagulants, there has been a paradigm shift in the use of anticoagulation to prevent stroke — as there is no longer a requirement for routine blood testing, as was the case with vitamin-K antagonist anticoagulants such as warfarin. Direct thrombin inhibitors such as dabigatran and FXa inhibitors such as rivaroxaban and apixaban have been developed and shown to be effective oral anticoagulants. Oral anticoagulants now also include the antiplatelet drug clopidogrel (Plavix).

Surgery

After Walton Lillehei's first success with open-heart surgery, Denton Cooley — a heart and cardiothoracic surgeon who, as mentioned earlier, would be the first to implant a totally artificial heart in a human patient — remarked, *"Lillehei had provided us with a 'can opener' to the biggest 'picnic' heart surgeons had ever known."* With the advances in heart surgery, some born with congenital heart disease such as Tetralogy of Fallot, who would not have survived infancy or childhood, are now living healthier lives into adolescence and adulthood; some of them are now even grandparents. Others with more complex problems can be helped considerably, while requiring closer follow up.

The era of progress in surgery for congenital heart disease and valvular heart disease was followed by another milestone: coronary artery bypass surgery. In our aging population, narrowing of the aortic valve (aortic stenosis) has now become a health concern starting at about age 65 but especially in persons age 85 and older. Many of these individuals have coexisting heart abnormalities, rendering these patients unsuitable for heart surgery. Transcatheter aortic valve replacement (TAVR) is showing encouraging results among that group. Former U.S. Secretary of State Henry Kissinger, another public figure, had coronary artery bypass surgery in 1988 and TAVR in 2014; as of the writing of this book, he is 97 years old.

Percutaneous angioplasty — introduced in 1979 by Andreas Gruentzig — has been, along with stent implantation, the procedure of choice in relieving coronary obstruction.

Interventional cardiology is fast encroaching upon the domain of cardiac surgery. Coronary artery bypass surgery is performed on the surface of the heart, and therefore, cardiopulmonary bypass is not essential. The results of simple beating-heart surgery are so satisfactory that with it, surgeons can now offer the patients the same degree of eventual comfort but better life expectancy compared to open-heart surgery. Using internal mammary artery in a minimally invasive procedure, rather than an interventional procedure, cardiac surgeons make smaller incisions on the right side of chest and, rather than cutting through the breast bone (a sternotomy), operate between the ribs —

resulting in less pain and quicker recovery. A combined approach is gaining increasing acceptance: using coronary artery bypass surgery for left anterior descending artery blockage, along with catheter-based coronary intervention such as angioplasty and stent implantation when other vessels are involved.

Toward a Wireless Heart

The term congestive heart failure refers to failure of the heart to maintain forward blood flow, resulting in lung congestion and symptoms such as shortness of breath and leg swelling. The treatment of end-stage congestive heart failure presently remains a major challenge.

The left ventricular assist device (LVAD), implanted in the heart, is a mechanical, pump-like device that takes the stress away from a failing heart by moving blood smoothly around the body without generating a pulse. The LVAD has become smaller and more reliable since its introduction in 1994. It was originally designed as a stop-gap measure to keep a patient alive for a few extra months while a suitable transplantable heart could be found. But improvement in the technology along with a shortage of donor hearts have resulted in the LVAD being increasingly used on a long-term basis.

One area in which the LVAD has not improved, however, is in power supply: The pump requires an electric cable, called a driveline, that runs through the abdominal wall to a battery pack in a harness. As the entire apparatus must be worn all the time, it is awkward for the wearer to take a shower and impossible for the person to swim — and driving a vehicle is discouraged because even a minor accident leading to dislodging of the driveline could be fatal. Most significantly, however, the LVAD is a constant source of infection. With most medical implants, such as pacemakers and hip replacements, the risk of infection decreases over time. But the LVAD driveline collects 'bugs' from the outside world, and the LVAD then introduces them directly into the bloodstream, allowing infection to spread quickly. As a result, the driveline must be redressed every day, using sterile gloves and gauze. Even so, virtually everyone fitted with a LVAD experiences an infection every 12 to 18 months. Many such infections are serious, and some are deadly.

Joshua Smith, an engineer at the University of Washington, and **Pramod Bonde**, a heart surgeon at the University of Pittsburgh, hope to change that. Between them they have developed the world's first wireless-powered, driveline-free heart pump. It was formally announced at the American Association for Thoracic Surgery's annual meeting in May 2012. Their Free-Range Resonant Electrical Energy Delivery system, or FREE-D, is powered by induction. Specifically, the FREE-D exploits a phenomenon called resonant coupling, in which metal coils that resonate at the same electrical frequency can exchange energy particularly efficiently. A transmitter coil, 26cm in diameter, can beam up to 15 watts of power to a receiver coil just 4.3cm across.

The transmitter coil can thus be worn in a vest that also holds a battery pack, while the receiver tucks neatly into the patient's chest. This allows efficient transfer of wireless power from an external transmitter to the implanted receiver. The next steps in research include refining the FREE-D system for animal trials. The wireless heart pump will not be available for a few years: Technical challenges remain, including integrating a leak-proof backup battery into the device and miniaturizing the FREE-D's control electronics. After those problems are resolved, the device will have to undergo extensive clinical trials in humans before it can be licensed for sale.

The improvements cannot come a moment too soon. The National Heart, Lung and Blood Institute estimates that more than 50,000 Americans a year need a heart transplant or an LVAD. Even so, fewer than 2,500 heart transplants are performed in the country every year due to donor shortages; and the number of LVAD implants, initially smaller, is now becoming more frequent. As the supply of donor hearts is unlikely to increase, the need for better LVAD technology is clear.

Prevention

On the prevention front: The Framingham Heart Study identified risk factors for developing coronary artery disease due to atherosclerosis, or 'hardening' of the arteries. Although considerable advances have been made in determining how atherosclerosis develops, the exact mechanism remains elusive.

However, attention to the cardiovascular risk factors has demonstrated dramatic benefits (just as with lung cancer, for which we do not know the exact cause but do know that cessation of smoking greatly reduces the risk).

Because of widespread prevention and with better treatment, deaths are being prevented among those with heart disease and primary events are being delayed. Life expectancy continues to increase as age-adjusted mortality tends to decline, with cardiovascular disease largely affecting older and still older individuals. The wise course of action is to modify cardiovascular risk factors before the disease develops or progresses. One is at a higher risk if one already has evidence of atherosclerotic diseases, for example: angina, heart attack, transient ischemic attacks or peripheral vascular disease.

Emotional/Mental Stress and Cardiovascular Disease

While temporary emotional or mental stress is often the result of worry, anxiety, impatience and anger, prolonged stress can have a harmful effect on our health and well-being. Although not a major risk factor for most people, ongoing stress may contribute to heart disease in some people. It may also contribute to other risk factors: For example, people under stress may overeat or start smoking or smoke more than they normally would. Stress of this kind involves the brain, mind, heart, and body; it involves our thoughts, emotions and feelings.

While we have learned much about this subject in recent times, the mind-body relationship has long been an area of interest in philosophy and literature, in science and medicine. In the 11th century, the physician Avicenna/ Ibn Sina drew attention, as noted earlier, to the primacy of intellect and the mind-body connection. Much later,

Dutch philosopher **Baruch Spinoza (1632-77),** in his masterwork *Ethics,* also made a major advance to our thinking in this area. He argued — correctly, as it turned out — that body and mind are not two separate entities but more of a continuous substance. (This countered the mind-body dualism thinking proposed earlier by the influential French philosopher, mathematician and scientist **René Descartes (1596-1650)**.)

In the late 16th-early 17th centuries, William Shakespeare wrote poems and plays that often depict the heart as the site of our most intense and secret feelings and thoughts — both virtuous ones, as in the heart of *King Lear*'s Cordelia (her name can mean "pure heart") and villainous ones, as in the heart of *Othello*'s Iago.

The heart's condition and activity and intense emotion are, without a doubt, closely interlinked and can have a great impact on one another. Mental stress can cast a long shadow on our future, particularly in the context of cardiovascular disease. In patients with clinically stable coronary artery disease, for example, mental stress-induced ischemia can be more common than exercise-induced ischemia.

Women, unmarried men and individuals living alone are at a higher risk of mental stress-induced ischemia. Psychological stressors appear to lead to microvascular dysfunction and metabolic issues that in turn serve as possible triggers for acute coronary syndromes. Such stressors can also have direct cardio-inhibitory effects leading to an acute cardiac disorder called Takotsubo syndrome or stress cardiomyopathy (also called broken-heart syndrome), which — while transient — may be recurrent and carries with it a substantial risk of mortality.

Medicine's challenge is to devise interventions so that lives are not cut short prematurely by the stress-cardiovascular disease relationship. There has been progress, but much remains to be done.

The persistence of modifiable risk factors that affect the large arteries and lead to heart attack or stroke, can also, we learned not long ago, clog small cerebral arteries — contributing to degenerative brain disease and acceleration of Alzheimer's disease. Magnetic resonance imaging (MRI) and sophisticated cognitive assessment testing have demonstrated that in patients with both conditions, especially those affected by cardiovascular risk factors, there is frequently blockage of small arteries and lacunar injury or micro-infarcts, while showing parallel decreases in cognitive ability.

Adherence to lifestyle changes and to medications/ prescriptions that aim to reduce modifiable cardiovascular risk factors would also reduce the risk of neurodegenerative disease, e.g., Alzheimer's disease. The prospect of not recognizing our loved ones or remembering what we did just a few minutes ago should make us as concerned as the prospect of having a heart attack or developing another cardiovascular condition.

How to Decrease the Risk of Developing Coronary Artery Disease

While coronary artery disease is still the leading cause of disability and death for both men and women, the considerable decline in its incidence is due largely to lifestyle and behavioral modifications — especially to less tobacco usage, prudent diet and moderate exercise.

Specifically, the risk of coronary artery disease can be decreased by:

- *Following a diet high in fruits and vegetables, and low in saturated fat, cholesterol and sodium*
- *Limiting food portion sizes*
- *Maintaining a healthy weight, including losing weight, if necessary*
- *Increasing physical activity, preferably exercising 30-60 minutes on most days of the week (at a moderately intense level)*
- *Not smoking or using tobacco in any form*
- *Routinely getting screenings for heart disease*
- *Taking medications for hypertension and hypercholesterolemia, as prescribed by a physician*
- *Taking medications for diabetes and obesity, as prescribed by a physician*

ii. Toward the Future

In the earlier part of the 21st century, our current era, many more interventions of all types (mechanical and electrical devices, cell and gene therapy and perhaps even transplants) will become increasingly available. The accompanying great expansion of the world's population, including many people developing diabetes and obesity, could also benefit from these interventions. There will be increasing sub-specialization and the role of the generalist, as the conductor of an orchestra, will become even more important.

With the digital revolution underway, important transformations will include improved communication between the patient and healthcare providers, a growing role for patients to manage their own health and have access to information available through electronic records, social media and large national databases. Soon, images from smartphone applications will come from CT scans and MRI images obtained from imaging centers, patient-generated echocardiography and electrocardiograms obtained from a small adapter and a smart watch that can measure blood pressure. The relentless growth of technology will enhance the world of medicine for the good of all.

Medicine might, additionally, benefit from a rollout of artificial intelligence. Algorithms that use machine learning applied to previously gathered data promise to predict whether someone has or will have some medical conditions such as heart failure.

Simultaneously, there will be a greater focus on prevention. The interventions will continue to be used as they become simpler, less expensive and more effective. However, with the application of genetics and genomics to heart disease, the need for intervention will decrease. The principal role of the cardiologist will change from recognizing and managing established disease to interpreting and applying genetic information in prevention and treatment.

Molecular medicine and genomics will profoundly influence the future in diagnosis and management of heart disease. There appear to be groundbreaking possibilities to significantly reduce or even essentially eliminate cardiovascular disease in the 21st century.

iii. Epilogue

The triumphant journey of great, grand achievements described in the preceding pages took place during the 20th century, which will therefore be remembered as the golden century of cardiology.

These innovations built upon basic research and the work over decades of numerous scientists, engineers and others. The advances came from collaborative efforts, for example, of cardiologists, engineers and physicists (electrocardiography and echocardiography), physicians and epidemiologists (the Framingham Heart Study) and scientists and industry (the first heart-lung machine, cardiac drugs and electronic devices).

Importantly, these were and are international triumphs: The work of researchers and investigators in numerous countries, on several continents, is described here. Others from many other nations have contributed significantly.

While these words are written as an epilogue to this book, the past is prologue: Let us always appreciate how the pioneers' endeavors and accomplishments have led us to today and are leading us toward tomorrow. Carried out in the interest of expanding human knowledge and in the service of humanity, the pioneering work and innovations have powered the march of scientific and medical progress in cardiology. Thanks to the pioneers and others, the march goes on, and the triumphs will continue. ◆

Acknowledgments

I would like to sincerely thank, by name, several people who assisted me in developing this book, either directly or through their influence and encouragement.

Yasmeen Haider, a physician as well as my wife and life partner, provided immeasurable and invaluable assistance and support from start to finish. I am extremely grateful to her.

My dear friend (the late) Ayub Khan Ommaya, MD, ScD, long encouraged me to write a book on cardiology. I am most thankful to him. A world-renowned neurosurgeon, he was also an illustrious researcher, inventor of the Ommaya reservoir, developer of a scoring system for traumatic brain injury, and himself a writer.

Karin Wingo, former manager of my medical practice, kindly and helpfully made many relevant informational sources readily available to me.

Sherry Migdail, an educator and philanthropist, and Sandra El-Khodary, a longtime family friend, conducted a general review of the book and provided numerous very useful suggestions.

Paul Wonnacott, a distinguished economist and later also a novelist, offered probing questions, a very helpful critique and excellent suggestions to make the book more easily understood by the general reader.

Saira Moini, a professional writer, editor and marketing/communications specialist, most expertly copy edited the manuscript's iterations and worked with great skill on the book's design, layout and publishing processes.

Nancy Dillon, a copyright specialist generously provided her expertise in research into photo sources and to obtain photo credits.

Grace Mott, diligently used her best efforts in obtaining photo credits and editing some changes.

Nadya Scheiner, an attorney, teacher and artist, as well as the eldest of our three daughters, created the artwork for the book's cover. Our daughters Aliya Haider, an attorney, writer and human rights activist, and Saira Haider, a program manager and business policy analyst, offered very welcome encouragement throughout the book project.

I extend my great thanks and deep gratitude to each of these individuals, among others.

I would also like to acknowledge and express sincere appreciation to and for the sources of information used in developing this book. The majority of these sources are listed in the bibliography at the end of the book. In addition, the book includes photos and images from the U.S. National Library of Medicine and from various websites, publications, and so forth, obtained via web research; I greatly appreciate the ability to use these.

Glossary

Anastomosing: Surgical joining of two blood vessels to form an intercommunication

Angina: Chest pain due to coronary artery disease

Angiogram: Procedure that provides detailed x-ray pictures of the heart and blood vessels

Angiography: Diagnostic or therapeutic radiography of the heart and blood vessels using a radio-opaque contrast material

Angioplasty: Endovascular procedure that reopens narrow blood vessels and restores forward blood flow

Angiotensin: Family of peptides that causes constriction of blood vessels and the blood pressure to rise

Aorta: Main branch of the body's arterial system arising from the heart, connecting heart and body

Arrhythmia: Disorder of the heart's electrical system resulting in a very rapid or very slow heartbeat

Asystole: Stoppage or cessation of beating of the heart

Atrial septal defect: Congenital heart abnormality, or opening, in the muscular wall (septum) that separates the right and left atria, which are the heart's upper chambers

B

Bradycardia: Slowing of the beating of the heart

C

Capillaries: Minute blood vessels that join the ends of the arteries to the beginnings of small veins and venules (through the capillary walls, oxygen diffuses to the body's tissues and products of metabolic activity enter the bloodstream)

Cardiac arrest: Sudden cessation of the body's functional blood circulation due to cessation of heart function. (Note: Cardiac arrest is an electrical problem triggered by disruption of the heart's rhythm, whereas a "heart attack" is a circulation problem. A heart attack may lead to cardiac arrest. Heart failure occurs when the heart muscle fails to pump as much blood as the body needs; it is usually a long-term, chronic condition, but may come on suddenly)

Cardiac defibrillation: Termination of ventricular fibrillation with electric counter-shock or spontaneous contraction of muscle fibers

Cardiopulmonary: Relating to the heart (cardio-) and the lungs (-pulmonary)

Cardiopulmonary resuscitation – see "CPR"

Cardiovascular: Relating to the heart (cardio-) and the blood vessels (-vascular)

Cardioversion: Conversion of an irregular heartbeat to regular rhythm by the application of a small dose of an electrical shock

Catheterization: Percutaneous use of a hollow tube, made of rubber or plastics, introduced through a vein in the arm and passed into the heart or great vessels for diagnosis or interventional treatment

Cerebral embolism: Sudden blockage of a blood vessel in the brain, e.g., by a blood clot

Commissurotomy: Incisions of the mitral valve's commissures to treat mitral stenosis (constricted valve) by increasing the size of the mitral orifice

Coronary artery: Blood vessel that supplies the heart muscle with oxygenated blood (rich in oxygen)

Coronary stent: Device used to prevent lumen closure and maintain open blood vessels after coronary bypass surgery and angioplasty

CPR (cardiopulmonary resuscitation): Resuscitation of normal heart (cardio-) and lung (-pulmonary) function

D

Defibrillator: Device that delivers an electric shock to the atrium or ventricle, with the goal of returning the electrical system to normal rhythm

Drug-eluting stents: Stents with a polymer coating over the mesh that allows a drug to be emitted over time, to prevent a blockage from recurring

E

Echocardiogram: Graphic record (sonogram) prepared by echocardiography that is used to visualize cardiac structures

Echocardiography: Use of ultrasound waves to investigate the action of the heart

Electrocardiogram: Graphic record prepared by electrocardiography

Electrocardiography: Creation and study of the graphic record (electrocardiogram) produced by electric currents originating in the heart

Electrophysiologist: Cardiologist who specializes in evaluating and treating heart rhythm abnormalities

Electrophysiology: Field of study of relationships of body functions to electrical phenomena and therapeutic use of electric currents

External cardiac pacemaker: Device located outside the body that can trigger mechanical contraction of the heart by emitting electric discharges transmitted through the chest wall

Extracardiac surgery: Surgery outside the heart

Extracorporeal: Outside (extra-) the body (-corporeal)

F

Fibrillation: A rapid and irregular heart rhythm caused by abnormal electrical impulses (this is a potentially serious condition, as due to these uncoordinated contractions of individual heart muscle fibers, the heart chamber involved cannot contract all at once and therefore pumps blood ineffectively, if at all).

G

Gastrocnemius muscle: Large muscle of the lower leg's posterior (the calf area)

H

Heart failure: Condition in which the heart's pumping function is inadequate to supply blood to the body, and so forth

Hypertension: Abnormally high blood pressure

I

Iatrogenic: Relating to illness caused by medical examination or treatment

Infarct: Area of cellular tissue (e.g., of the heart or the brain) that undergoes necrosis, or death, following insufficient blood supply

Infarction: Death of living tissue (myocardial infarction, or heart attack; cerebral infarction, or stroke)

Interventional cardiology: Sub-specialty of cardiology that uses intravascular catheter-based techniques with fluoroscopy to treat heart abnormalities (coronary artery, valvular or congenital heart disease)

Interventricular septum: Wall dividing the two ventricular cavities, or ventricles, of the heart

Intracardiac repair: Repair within the heart

Intravascular: Situated or occurring within (intra-) a blood vessel (-vascular) or blood vascular system

Ischemia: Temporary deficiency of blood flow (and therefore, of oxygen) to an organ or a tissue

L

Lumen: Canal within a tubal structure of the body, e.g., the coronary artery lumen

M

Mitral regurgitation: Backward blood flow through a defective mitral valve

Mitral (valve) stenosis: Narrowing or constriction of the heart's mitral valve

Mitral valve: One-way valve allowing blood to flow from the heart's left atrium to the heart's left ventricle

Myocardial: Relating to the muscle (myo-) that makes up the heart (-cardial)

Myocardial infarction: Heart attack

O **Orthotopic:** Cardiac transplant in the correct place, i.e., to a site where that tissue would normally be present

P **Pacemaker:** Specialized tissue or device that governs a rhythmic biologic activity. The heart possesses a natural pacemaker that can, if faulty, be replaced by an electronic device

Patent ductus arteriosus: Abnormal persistence after birth of a communication between the main pulmonary artery and the aorta

Percutaneous: Done through (per-) the skin (-cutaneous)

Percutaneous angioplasty: Catheter-attached device introduced through the skin to open up narrowed or blocked blood vessels

Percutaneous AVR: Aortic valve replacement (AVR) performed by introducing a catheter through the skin

Pericardial effusion: Fluid accumulation in the pericardium surrounding the heart

Pericardium: Membrane that encloses the heart, consisting of an outer fibrous layer and an inner double layer of serous membrane

Pulmonary: Relating to the lungs

R **Revascularization:** Restoration of blood flow

S **Sciatic nerve:** Largest nerve in the body, arising from sacral plexus on either side of body and passing from pelvis to back of the leg

Statin: Class of drugs that lowers blood cholesterol; the drugs are also known as HMG-CoA reductase inhibitors.

Stenosis: Constriction or narrowing of a passage or orifice

Stent: Device used to prevent lumen closure and to maintain open blood vessels

Sternotomy: Operation of cutting through the sternum, the breastbone (chest bone) that forms the front of the rib cage

Streptokinase: Clot-busting agent that helps to remove thrombotic clots from the blood vessels

T **Tetralogy of Fallot:** Congenital malformation of the heart consisting of four defects or abnormalities: pulmonary valve stenosis, right ventricular hypertrophy, ventricular septal defect and displaced aorta (malformation was named for French physician Étienne-Louis Arthur Fallot (1850-1911))

Thorax: Chest

Thrombolytic: Causing the breakup of a blood clot

Transaminase: Enzyme released by damaged cells; a high serum level of this enzyme may be diagnostic of myocardial infarction, or heart attack

Transfemoral: Done through the femoral vessels – the artery or vein

V

Venous: Relating to the veins

Ventricle: One of the heart's two lower chambers; when filled with blood, the ventricles contract to propel the blood to the arteries. The right ventricle pumps deoxygenated venous blood to the lungs (to get re-oxygenated); the left ventricle pumps oxygenated blood to the body

Ventricular fibrillation: Quivering or spontaneous contraction of muscle fibers; a lethal dysrhythmia resulting in the clinical absence of effective circulation of the blood (pulse-lessness)

Ventricular septal defect (VSD): Hole in the wall between the heart's right and the left ventricles; may be accompanied by dizziness or fainting or even result in sudden death

Bibliography

PROLOGUE

Aristotle. (2015). *On the Heavens (De caelo)* translated by J.L. Stocks. University of Adelaide Press.

Darwin, C. (1952, 1982). *The Origin of Species by Means of Natural Selection*. Encyclopaedia Britannica, Inc.

Hawking, S.W. (1988). *A Brief History of Time*. London: Bantam Press/Books.

Harper, R.F. (1904). *The Code of Hammurabi, King of Babylon about 2500 BC*. Chicago: University of Chicago Press.

Harvey, W. (1847). *The Works*, Tr. by R. Willis, v.10, p.597. London: Sydenham Society Publication.

Harvey, W. (1938). *On the Motion of the Heart and Blood in Animals*. Tr. by R. Willis, Harvard Classics, Collier, v.38, pp.75-149.

Harvey, W. (1628). *Exercitatio anatomica de motu cordis e sanguinis in animalibus* (*An anatomical disquisition on the motion of the heart and blood in animals*), London. (1) Tr. by R. Willis, Surrey, England: Barnes, 1847. (2) English translation with annotations by C.D. Leake, Springfield, Illinois: Charles C. Thomas.

Iskandar, A.Z. (1967). *A Catalogue of Arabic Manuscripts on Medicine and Science*. London: Wellcome Historical Medical Library.

Meyerhof, M. *Iban-an-Nafis und seine, Theorid des hungenkreislaufs. Quell, u, Stud. Gesch, der Nuturwiss. u. der. Med.*, 1933:4, 37.

Newton, Sir Isaac. (1999.) *The Principia: Mathematical Principles of Natural Philosophy*. Tr. by A. Motte. Rev. by F. Cajori, University of California Press. Encyclopedia Britannica, Inc.

Osler W. (1908.) *Biographical Essays*. Harvey W. Howe, Brydone, Printers Ltd.

Porter, R. (1997). *The Greatest Benefit to Mankind: A Medical History of Humanity*, W.W. Norton & Co.

Silverman, M.E. "William Harvey and the discovery of circulation." Clin. Cardiol, 1985:8, 244-6.

ELECTROCARDIOGRAM

Barker, LF. Electrocardiography and phonocardiography: a collective review. Bulletin of Johns Hopkins Hospital, 1910:21, 355-59.

Besterman, E. (1979). Waller- pioneer of electrocardiography. Br. Heart J., 42, 61-4.

Burch, G.E., DePasquale, M.P. (1964). *A history of electrocardiography*. Chicago: Year Book Medical Publishers.

Burnett, J. The origins of electrocardiograph as a clinical instrument. Med Hist. Suppl. 1985:5; 53 [NLM - PubMed].

Snellen H.A. (1995). *Willem Einthoven (1860–1927) – Father of electrocardiography. Life and work, ancestors and contemporaries.* Dordrecht/Boston/London: Kluwer Academic Publishers, pp. 5-60.

Eshler, I. "Willem Einthoven - the man. The string galvanometer electrocardiograph." Arch. Int. Med., 1988:148, 453-5.

Flemming, J.A. (1910) *Electricity*. In: Encyclopaedia Britannica - 11th ed., vol. 9, 1979.

Galvani, L. *Aloysii Galvani de vibruselectricitatisin motu muscularis. De Bononiensi Scientiarum et Atrium Instituto atque Acadamia, cmentarii* 1791; 7: 363-418.

Gilbert, W. (1600). *De magnete, magnetique corporitus, et. magnomagnete tellure; physologia nova, plurimio and augmantis and experimentis demonstrate*. London: Short.

Gilbert, W. (1900). *On the magnet, Magnetick Bodies also on the Great Magnet, the Earth,* tr. by S.P. Thompson. London: Cheswick.

Hill, AV. An obituary of Einthoven. Nature, 1927:120, 591.

Kollicker, A. and Muller, H. *Nachiveis der negative, Schwankung des Muskelstromes am natrulich sich contrahirenden. Muskel. Verb. Phys. Ges,* 6, 528-33.

Lama, A. "Einthoven: the man and his invention"(in Spanish). Rev. med. Chil, 2004; 132 (2): 260-4.

Lewis, T. (1913). *Clinical Electrocardiography*. New York: Shaw and Sons, p.11.

Nobel lectures, physiology or medicine, 1922-1941. (1965). Amsterdam: Elsevier Publishing.

Piccolino, M. Animal electricity and birth of electrophysiology the legacy Luigi Galvani. Brain Res Bull, 1998;46(5);381-407.

Piccolino, M. Luigi Galvani's path to animal electricity. CR Biol. 2006;329(5-6):303-18.

Snellen H.A. (1995). *Willem Einthoven (1860-1927), Father of electrocardiography: Life and work, ancestors and contemporaries.* Springer.

Snellen, H.A. (1977). *Selected papers on electrocardiography of Willem Einthoven.* The Hague (Netherlands): Leiden University Press.

Waller, A.D. (1922) *The Electrical Action of the Human Heart.* University of London Press.

Waller, AD. A demonstration on man of electromotive changes accompanying the hearts beat. J. Physiol, 1887;8(5):229-34.

CARDIAC CATHETERIZATION

Abrams, H.L. Angiocardiography and thoracic aortography. In: Bruwer, A.J., ed. (1964). *Classic descriptions in diagnostic oentgenology.* vol. 1. Springfield (IL): Charles C. Thomas, pp.492-501.

Buizzi, A. Claude Bernard on cardiac catheterization. Am. J. Cardiol., 1954:4,405.

Cournand A, Ranges HA. Catheterization of the right auricle in man. Proc. Soc. Exp. Biol. Med., 1941:46; 24, 462.

Cournand, A. Cardiac catheterization. Acta. Med. Scand. Suppl., 1975:579.

Editorial. Stephen Hales: Father of hemodynamics. Med. Times, 1944:72, 315.

Forssmann, W. *Die Sondierung des Rechten Herzens. Klin. Wochnschz.* 1929:8. 2085-7.

Forssmann, W. (1974). *Experiments on myself. Memoirs of a surgeon in Germany.* New York: St. Martin's Press.

Hales, S. (1733). *Statical Essays: Containing hemodynamics.* London: Innys and Mandy.

Lequime, J. Andre Cournand (1985-1988). His role in development of modern cardiology. Acta Cardiol. 1988:XL111-(4);437-42.

McMichael J. and Scharpey-Schafer EP. The action of intravenous digoxin in man. Q. J. Med. 1944: 13, 1123.

Mueller, RL, Sanborn, TA. The history of interventional cardiology: cardiac catheterization, angioplasty and related interventions. Am. Heart J. 1995; 129: 146-72.

Richards, DW. Cardiac output by catheterization technique in various chemical conditions. Fed. Proc. 1945; 4; 215-20.

Sourkes, T.L. (1966). *Nobel Prize Winners in Medicine and Physiology 1901-1965.* New York: Abelard Schuman.

CORONARY ANGIOGRAPHY

Monagan, D.D. and Williams, D.O. (2007). *Journey into the Heart: A Tale of Pioneering Doctors and Their Race to Transform Cardiovascular Medicine.* New York: Gotham.

Meyers, M.A. (2007). *Happy Accidents: Serendipity in Modern Medical Breakthroughs.* New York: Arcade Publishing.

Sheldon, W.C. (2008). *Pathfinders of the Heart. The History of Cardiology at the Cleveland Clinic.* Bloomington, IN: Xlibris Corp., pp. 26-7.

Sones FM, Jr., Shirey EK. Cine coronary angiography. Mod. Concepts Cardiovasc, Dis. 1962;31: 735-8.

Vineberg, A. Surgery of Coronary Artery Disease. Prog. Cardiovasc. Dis. 1962:391-418.

CARDIOVASCULAR SURGERY and MINIMALLY INVASIVE PROCEDURES

Bernard C.N., Curtis B.P. (1969). *One Life.* New York: Macmillan.

Bernard CN. A human cardiac transplant: An interim report of a successful operation performed at Groot Schoor Hospital, Capetown. S Afr Medical J. 1967; 41:1271.

Blalock A, Taussig HB. The surgical treatment of malformations of the heart in which there is pulmonary stenosis or pulmonary atresia. JAMA. 1945;128(3):189-202.

Carrel, A. (1902). *La Technique Operatoire des Anastomoses et la Transportation des Visceres, Lym Medical* (The operative technique of vascular anastomosis and transplantation of organs). In: Hurwitz, A. and Degenshein, G.A. (1958). *Milestones in Modern Surgery* New York: Paul B. Hober, Inc., pp.374-9.

Cooley, D. In Memoriam: C. Walton Lillehei, the 'Father of Open-Heart Surgery.' Circulation (AHA Journal), 1999;100:1364-1365.

Eloesser, L. Milestones in chest surgery. J Thoracic and cardiovascular Surg, 1970; 60;157-65. [NLM - PubMed].

Favaloro, RG. The present era of cardiovascular revascularization some historical landmarks. Int. J. Cardiol., 1983:4, 331-44.

Favaloro RG, Effler DB, Cheanvechai C, Quint RA, Sones FM, Jr. Acute coronary insufficiency (impending acute myocardial infarction and myocardial infarction); Surgical treatment by the saphenous vein graft technique. Am. J. Cardiol. 1971; 28:598-607. [NLM - PubMed].

Favaloro RG. Saphenous vein autograft replacement of severe coronary artery occlusion. Ann Thorac Surgery. 1968; 5:334.

Gharagozloo, Farid and Najam, Farzad. (2008). *Robotic Surgery.* McGraw-Hill Professional.

Gibbon JH, Jr. Application of a mechanical heart and lung apparatus to cardiovascular surgery. Min. Med. 1954;37:171-5.

Gross RE, Hubbard JH, Surgical ligation of a patent ductus arteriosus: report of first successful case. JAMA, 1939:112;729-33.

Haider R, Finnegan P, Sanchos P, Singh SP, Abrams LD, and Parsons D. Twelve-year follow-up of Tetralogy of Fallot after surgical Correction of Fallots Tetralogy in Childhood. Brit. H.J. 1972:34,2,205,

Harken DE, The emergence of cardiac surgery. Personal recollections of 1940's and 1950's. J Thorac Cardiovasc Surg. 1989:98(5 Pt 2):805-813

Harken DE. Foreign bodies in and in relation to, the thoracic blood vessels and heart; techniques for approaching and removing foreign bodies from the chambers of the heart. Surgic Gynecol Obstet. July 1946:83;117-125.

Lillihei CW, Gott VL, DeWall RA, Varco RL. The surgical treatment of stenotic or regurgitant lesions of mitral and aortic valves by direct vision utilizing pump oxygenator. J Thoracic Surgery 1958:35;154-90.

Lower, RR and Shumway, NE. Studies in ortho-homotransplantations, Surg. Forum, 1960:1,18.

McNamara, DG. The Blalock-Taussig operation and the subsequent progress in the surgical treatment of cardiovascular disease. JAMA, 1984:251, 2139-41.

Monagan, D.D. and Williams, D.D. (2007). *Journey into the Heart: A Tale of Pioneering Doctors and Their Race to Transform Cardiovascular Medicine*. New York: Gotham.

Rehn L, On penetrating cardiac injuries and cardiac suturing. Arch Kline Chir. 1897: 55: 315.

Richmond, C. "Norman Shumway." BMJ. 2006, Mar 4: 332(7540): 553. [PMC: U.S. National Library of Medicine, NIH].

Shumacker, Jr., H.B. (1999). *A Dream of the Heart: The Life of John H. Gibbon, Jr., Father of the Heart-Lung Machine*. Fithian Press.

Taussig, HB. On the evolution of our knowledge of congenital malformation of the heart, the T. Duckett Jones Memorial Lecture. Circulation (AHA Journal), 1965:31,768-77.

Thomas, V.T. (1985). *Pioneering Research in Surgical Shock and Cardiovascular Surgery: Vivien Thomas and His Work with Alfred Blalock – an Autobiography*. Philadelphia: University of Pennsylvania Press.

CORONARY CARE UNIT

Beaudouin, D. "W.B. Kouwenhoven: Reviving the Body Electric." *JHU Engineering* magazine, Fall 2002.

Braunwald, E. Thirty-five years of progress in cardiovascular research. Circulation., 1984;70(5 pt 2);1118-25. [NLM - PubMed].

Brown K.W.G., McMillan, R.L. The effectiveness of the system of coronary care. In: Meltzer, L.E. and Dunning A.J, eds. (1972). *Textbook of coronary care*. Amsterdam: Excerpta Medica, pp.32-7.

Day, HW. A cardiac resuscitation program. The Lancet, 1962:82:153-156.

Julian DG. The history of coronary care units. Br. Heart J. 1985:7:497-502.

Julian, DG. Treatment of cardiac arrest in acute myocardial ischemia and infarction. The Lancet, 1961:ii: 840-844.

Killip T, 3rd, Kimball JT, Treatment of myocardial infarction in a coronary care unit. A two-year experience in 250 patients. Am. J. Cardiol. 1967:20;457-64.

Kouwenhoven WB, Knickerbocker GG. Closed-Chest cardiac massage. J Am Med Assoc. 1960:173(10) 1064-1067.

Lown B, Fakhro AM, Hood WB. Jr., Thorn GW. The coronary care units: New perspectives and directions. JAMA, 1967; 199:188-98. [NLM - PubMed].

Meltzer, L.E. Presbyterian-University of Pennsylvania Medical Center. Philadelphia. In: Julian, D.G., Oliver, M.F., eds. (1968). *Acute Myocardial Infarction*. Edinburgh: Livingstone. pp.3-6.

Zoll P, Paul MH, Lilenthal AJ, Gibson W, Paul MH, Norman LR. Termination of ventricular fibrillation in man by externally applied electric countershock. N Engl J Med. 1956: 254:727-732.

PREVENTIVE CARDIOLOGY

Brink, S. Unlocking the heart's secrets. *U.S News & World Report*, 1998:125: 56-99

Haider R, Ahmad G. Health and Longevity in Hunza. P.J.M.R. 1966:V-2: 133-140.

Kannel WB, Dawber TR, Kagan A, Revotski N, Stokes J, 3rd. Factors of risk in the development of coronary heart disease: Six-year follow up experience. Ann. Int. Med. 1961:55; 33-50.

Naughton, J. & Haider, R. Methods of Exercise Testing. In: *Exercise Testing and Exercise Training in Coronary Heart Disease*. (1973). Eds., Naughton, J. and Hallerstein, H. New York: Academic Press, pp.71-91.

White, P.D., ed. (1944). *Heart Disease.* 3rd ed. New York, NY: McMillan Co., pp.120-25.

CARDIOVASCULAR DRUGS

Black JW, Stevenson JS, Pharmacology of a new adrenergic beta receptor compound. The Lancet 1962:2; 311-4.

Brown MS, Goldstein JL. A receptor mediated pathway for cholesterol homeostasis. Science, 1986:232; 34-41.

Endo, A. The discovery and development of HMG-CoA inhibitors. J. Lipid Res 1992:33: 1569-82. [NLM - PubMed].

Endo, A. A historical perspective on the discovery of statins. Proc Jpn Acad Ser B Phys Biol Sci. 2010, May 11; 86(5):484–493.

Goldstein JL, Brown MS. Cholesterol Up Close: Cholesterol – a Century of Research. HHMI Bulletin, Sept. 2003.

Sherry S, Fletcher AP, Alkjaezsig N, Smyzniotis FE. An approach to intravascular fibrinolysis in man. Trans. Assoc. Am. Phys 1957:70: 288-96.

Steinberg D, Gotto AM, Jr. Preventing coronary artery disease by lowering cholesterol levels: fifty years from bench to bedside. JAMA 1999:282; 2043-50. [NLM - PubMed].

ECHOCARDIOGRAPHY

Bommer WJ, Miller LR, Mason DT, and DeMaria AN. Enhancement of pulsed Electrocardiography in the evaluation of aortic valve disease by development of computerized spectral frequency analysis. Circulation (AHA), 1978: (suppl 11) Abstr. 58,187.

Burckhardt, CB. Comparison between spectrum and time interval histogram of ultrasound Doppler signals. Ultrasound Med. Biol., 1981:7,79.

Curie P, Curie J. *Development, per pression de l'ectricite polaire dans les cristaux hemiedres a faces inclinees. Comptes Rendus.* 1880; 91: 291-5.

Edler, I. The diagnostic use of ultrasound in heart disease. Acta Med. Scand. Suppl., 1955:08,32.

Edler I, Hertz CH. Use of ultrasonic electroscope for the continuous recording of movements of heart wall. Kungl Fysiogr Salbk Lund Forth, 1954:24-40.

Edler I, Hertz CH. Ultrasound cardiogram in mitral valve disease. Acta Ch. Scand.,1956:111,230.

Feigenbaum, H. Evolution of echocardiography. Circulation (AHA), 1996:93:1321-7 [NLM - PubMed].

Feigenbaum, H. History of echocardiography (monograph on the Internet; cited 2005). (http://www.asecho.org/freepolf/FeigenbaumChapter.pdf)

Joyner CR, Reid JM, Bond, JP. Reflected ultrasound in the assessment of mitral valve disease. Circulation (AHA), 1963: 27:503-11. [NLM - PubMed].

Nilson J, Westling H. Ultrasound in Lund—Three world premieres. Clin Physiol Funet Imaging, 2004; 24: 137-40.

Westling, H. "Inge Edler and Hellmuth Hertz and their pioneering work with ultrasound in the diagnosis of diseases of the heart." Department of Internal Medicine, Faculty of Medicine, Lund University, 29 October 1953. (https://www.med.lu.se/english/about_the_faculty/history_of_the_faculty/personalities_discoveries_and_innovations/inge_edler)

INTERVENTIONAL CARDIOLOGY

Dotter, CT, Judkins MP. Transluminal treatment of arteriosclerotic obstruction -description of a new technic [technique] and a preliminary report of its application. Circulation, 1964, Nov 30:654-70.

Dotter, CT, Transluminal angioplasty: A long view. Radiology, 1980 June; 135(3):561-4.

Gruentzig AR, Myler RK, Hanna ES, Turina, MI, Coronary transluminal angioplasty (abstr). Circulation (AHA), 1977:84: 55-6.

Gruentzig AR, Senning A, SiergenthalerWE, Nonoperative dilatation of coronary-artery stenosis: percutaneous transluminal coronary angioplasty. N. Engl. J. Med. 1979:301: 61-8.

Hurst, JW. The first coronary angioplasty as described by Andreas Gruentzig. Am J Cardiol. 1986; 57:185-6. [NLM - PubMed]

Mueller RL, Sanborn TA. The history of interventional cardiology: cardiac catheterization, angioplasty, and related interventions. Am. Heart J. 1995:129: 146-72.

PACEMAKERS AND DEFIBRILLATORS

Elmqvist, R. Review of early pacemaker development. Pacing Clin. Electrophysiol. 1978:1: 535-6.

Elmqvist, R, Senning, A. Implantable pacemaker for the heart. In: Smyth, CN, ed. (1960). *Medical electronics: Proceedings of the second International Conference on Medical Electronics*, Paris, June 1959. London: Illife and Son.

Haider R, Meyer JF, Rasul A. Cardiac Pacemakers: Current Concepts. Amer. Family Physician, May 1984:223

Mirowski M, Mower MM, Staeuben WS, Tabatznick B, Mendeloff AI. An approach to prevention of sudden coronary death. Arch. Inter. Med. 1970:26: 158-61.

Mirowski M, Reid PR, Mower MM. Termination of malignant ventricular arrhythmias with an implanted automatic defibrillator in human beings. N. Eng. J. Med. 1980:303: 322-4.

Mulpuru, SK, et al. Paradigm shifts in cardiac pacemakers. J Am Coll Cardiol. 2017:69(2): 189-210.

Zoll, PM. Resuscitation of the heart in ventricular standstill by external electric stimulation. N. Engl. J. Med. 1952:247: 768- 71.

Index

Images

Image Credits

The author would like to thank these individuals and institutions listed for permission to reproduce copyright material. While every effort has been made to trace all copyright holders, he wishes to apologize should there have been any errors or omissions.

William Harvey, Credit: Wellcome Collection

A.D. Waller, Credit: Wellcome Collection

Willem Einthoven, Credit: Alamy

ECG graphs, Academic Medical Center of the University of Amsterdam. Courtesy of Dr. Jonas de Jong.

Coronary Angiography, Credit: BruceBlaus

Andre Cournand, Archives and Special Collections of Columbia University. Courtesy of Stephen Novak.

First selective coronary arteriogram, Courtesy of the Michael Schwartz Library at Cleveland State University.

Mason Sones, Courtesy of the Michael Schwartz Library at Cleveland State University.

Atlas of Congenital Cardiac Disease, McGill University. Special thanks to Andrew Murray.

Dwight Harken, Courtesy of the U.S. National Library of Medicine.

Willem Einthoven, Courtesy of the U.S. National Library of Medicine.

Vivien Thomas, Alan Mason Chesney Medical Archives at Johns Hopkins University.

Alfred Blalock, Estate of Yousuf Karsh. Special thanks to Julie Grahame.

C. Walton Lillehei, Courtesy of University of Minnesota Archives, University of Minnesota- Twin Cities

John and Mary Gibbon, Courtesy of Thomas Jefferson University, Archives and Special Collections. Special thanks to Kelsey Duinkerken.

Rene Favoloro and Mason Sones, The Cleveland Clinic Center for Medical Art & Photography © 2007 All Rights Reserved.

Norman Shumway, Courtesy of Stanford University Archives and Special Collections. Special thanks to Pamela Moreland.

Frank Pantridge, Courtesy of Pacemaker Press. Special thanks to David McCormick.

E.I. Chazov, Credit: Yuri Golovin

Michael Brown and Joseph Goldstein, Courtesy of Dr. Brown and Dr. Goldstein.

Senning, Elmqvist, & Crafoord, Credit: Professor Marko Turina, University Hospital, Zurich

About the Author

Riaz Haider, MD, FACP, FACC, FRCP, a physician, cardiologist and medical educator, is now retired from active medical practice.

During a career spanning 5 decades, Dr. Haider, a former Clinical Professor of Medicine at the George Washington University School of Medicine and Health Sciences (1984-2011) and President of the American Heart Association – the Nation's Capital Affiliate (1990-92), served as Chief of Cardiology at the Providence Hospital in Washington, DC (1974-93). He was also President of Washington Cardiology Associates, PC (1974-2008).

In the earlier part of his career, he held the positions of Senior Medical Registrar (Cardiology) at the Hammersmith Hospital and Tutor in Medicine at the Imperial College of Medicine, London (1971-72).

Dr. Haider is an elected Fellow of the American College of Physicians (ACP), American College of Cardiology (ACC) and the Royal College of Physicians (RCP), Edinburgh.

Honors received include: The American Heart Association, Nation's Capital Affiliate – Heart of Gold Award, 1993; Providence Hospital – Gold-Headed Cane Award, 1992; and American College of Cardiology – Member, Order of William Harvey, 1977.

Made in the USA
Monee, IL
08 November 2026

67377689R00121